Wallase

SHROPSHIRE CLOCK AND WATCHMAKERS

By the same author:
Buckingham: the Loyal and Ancient Borough

SHROPSHIRE CLOCK AND WATCHMAKERS

by

DOUGLAS J. ELLIOTT

PHILLIMORE

1979

Published by
PHILLIMORE & CO. LTD.,
London and Chichester

Head Office: Shopwyke Hall, Chichester,
Sussex, England

© Douglas J. Elliott, 1979

ISBN 0 85033 328 8

Printed in England by
UNWIN BROTHERS, LTD.,
at the Gresham Press, Old Woking, Surrey

and bound at
THE NEWDIGATE PRESS, LTD.,
at Book House, Dorking, Surrey

To My Wife
Audrey

A SHROPSHIRE WATCHMAKER

Thy movements, Isaac, kept in play,
Thy wheels of life felt no decay
For fifty years at least;
Till by some sudden secret stroke,
The balance or the mainspring broke,
And all the movements ceas'd.

Epitaph of Isaac Wood, 1736–1801
Watchmaker of Shrewsbury

CONTENTS

	page
Preface	viii
Acknowledgements	x
Introductory Chapter	1
The Company of Smiths, Shrewsbury	18
The Company of Hammermen, Ludlow	20
The Worshipful Clockmakers' Company, London	23
Biographical List of Clock and Watchmakers	25
The Story of J. B. Joyce & Co., Ltd., Whitchurch	137
Thomas Vickery, Clockmaker Extraordinary	158
Appendix. Letters of John Briant, Clockmaker, 1812	161
Craftsmen by Towns and Villages	164
Addenda	172

LIST OF ILLUSTRATIONS
(between pages 50 and 51)

1.	William Peplow, 1794-1895
2.	John Trevor Griffith Joyce, 1901-1971
3.	Norman Joyce, 1891-1966
4, 5 & 6.	Advertisements in *Eddowes Journal*
7 & 8.	Advertisement in *Salopian Journal*
9.	Watchpapers
10, 11, 12 & 13.	Silver pair cases for Verge Watches, Thomas Gorsuch, Shrewsbury
14.	Long-case clock, Francis Arkenstall
15.	Long-case clock, T. Vickery
16.	Watchpaper, Shrewsbury
17.	Long-case clock, William Marston
18.	Long-case clock, John Baddiley
19.	The Tower, Town Walls, Shrewsbury

PREFACE

Contained within the pages of this slim volume are the details of approximately 540 clock and watchmakers who have plied their craft within the county of Shropshire. Compiled from many and varied sources the biographical list extends from the early 16th century up to the year 1875, the latter year being chosen as being a century previous to the year when the research for this book was begun.

It is not the intention of the author to engage in the old argument regarding as and when clockmakers ceased to be makers and became merely shopkeepers selling mass-produced clocks with their names inscribed upon them. Except for the early makers, very few completely made every part of their clocks; often parts were bought in to be assembled, wooden cases being made by local cabinet-makers.

An example of a Shropshire craftsman making parts for others of his trade is illustrated in an advertisement put out by Charles Blakeway (1749-1809) of Allbrighton:

> Wheels cut for clockmakers at 6^d per set and Dyal Plates engraved at 2s. 6d each.

Even the most famous of clockmakers, Thomas Tompion is reputed to have had a factory producing rough movements which were finished to each customer's requirements, for no single man could have produced so many pieces in a lifetime. About 6,000 items are known by him, excluding those which must have been destroyed in the 300 years since his time. So one cannot pinpoint any specific year as a halting place for a list of actual clockmakers.

Most people are interested in the particular clock in their possession, whether the man whose name appears on its face actually made each separate part, or merely assembled parts produced by another, is a technicality not really important to them, they just want to know the approximate age of the clock or watch and some details about the maker whose name appears upon it. In this book it is hoped that those with clocks and watches made by Shropshire craftsmen will be able to find the answers to their queries.

ACKNOWLEDGEMENTS

In the compilation of such a book as this the net of research has to be flung far and wide, to bring in all those odd items of information which total up to a worthwhile volume.

The author's grateful thanks are due to the many people who made this possible. To Mr. Anthony Carr, Local History Librarian of the Salop County Library, and his assistants, Miss Honor Williams and Mrs. May Ion, who over the entire period of my research have shown endless patience and co-operation in supplying my almost insatiable demands for books and documents. Mr. R. E. James, Curator of the Clive House Museum, Shrewsbury, for allowing me access to the collection of watches and watchpapers in his care. The Salop County Archivist and her staff for the production of many records for my inspection. Dr. John Frost, Librarian, General University Library, New York, for information relating to the Joyce family in New York. Miss J. A. Hughes for her kind encouragement and helpful information relating to sales of Shropshire clocks within recent years. Mr. L. A. Burman, Keeper of Decorative Arts, Merseyside Museums, for information and photographs relating to the Gorsuch watch in Liverpool Museum. The County Archivist, Clywd, and his staff, for extracts from the Census Returns referring to the Joyce family in Denbighshire. Mrs. Margaret Joyce of Whitchurch for the very kind loan of family papers. Mr. Cyril Joyce of Corwen for information relating to his branch of the family of Joyce. The National Library of Wales for supplying photostats of wills of Shropshire clock and watchmakers.

Last, but not least, to my wife, who has so patiently and sympathetically listened to my seemingly endless talk about clockmakers.

INTRODUCTION

*'Tis with our judgements as our watches, none
Go just alike, yet each believes his own.*
—Alexander Pope

CLOCKS OF A BYGONE AGE today have a tremendous appeal. The great popularity of antiques of all kinds and their enormous appreciation in value over recent years have done much to enhance the ownership of these mechanical timekeepers.

The long-case clock with its slow, soothing 'tick-tock' is a special favourite with most people. Standing dignified against the wall and showing the passing of time on its face, it gives one the impression that it has always been standing there, timeless; truly a 'grandfather' of clocks.

Watches, too, have a fascination with their quiet but very busy ticking which makes them sound so very fussily important, as if it were urgent that they tick away the hours as fast as possible.

Old clocks and watches are both a great talking point, and this leads us to want to know more about the craftsmen who made them. Although it is not possible to delve too deeply into their lives, it is feasible from such varied sources as parish registers, advertisements in local newspapers and entries in directories, to gain an insight, not only of the men themselves, but also something of their families. Census Returns of the 19th century are even more helpful as they list the members of the family, their ages and their birthplaces. From these Returns it is possible to observe the movements of the tradesmen, for often their places of birth are far from the county of Shropshire, where they have finally settled in business. Many come from such widely-spaced places as the Orkney Islands; Frome, in Somerset; Coventry; London; Liverpool, and numerous other places.

One pictures this county acting like a lodestone, drawing these craftsmen from the four corners of the British Isles to settle here in Shropshire, but, of course, this was happening in other counties, too; just as we know that the process happened in reverse, with Shropshire-born craftsmen moving to other counties, and even to other countries. To mention just a few of the latter:

Job Rider (1761-1833), born at Westbury, who went to Ireland, and set up in business in the City of Belfast, where his fine workmanship, and character, earned him the highest of praise. Robert Joyce (1754-1798), born at Cockshutt, who left his father's business and travelled to London, where he mastered the intricacies of his trade, after which he too, like Job Rider, went to Ireland, but in his case to Dublin, from where his wanderlust took him across the ocean to America, to set up in business in New York. His brother, John Joyce (1744-1809) left his birthplace of Cockshutt to settle at Ruthin, just over the Welsh border, a business which his descendants moved to Denbigh, where the name of Joyce can still be seen over the local watchmaker's shop. Two other brothers, Samuel Joyce (1759-1827), and Conway Joyce (1761-1836) were attracted to London, where their business flourished in Lombard Street, and where they were honoured by a visit from George IV, who came to their shop to see the finely-made jewelled watch which they had made for the Emperor of China. These men and many others found distant places more attractive (financially) than their own birthplaces within this county of Shropshire; a see-saw movement, sometimes gaining craftsmen, sometimes losing them.

Who, when, or where the first clockmaker flourished in Shropshire is unknown; his name is hidden in the mists of time, that very time which he endeavoured to regulate. It is known that at Wenlock Priory there was a 'horlogium' in the reign of Henry III, for on the Close Rolls of that monarch, dated 1233, appears:

'De quercubus datis,—Mandatum est P. de Rivall' quod in forest a
regis de Shirlet faciat habere sacriste de Wenlac sex quercus ad

Introduction

horlogium suum apud Wenlak' faciendum de dono regis. Teste rege apud Wenlak'. vj die Junii. Per ipsum regem coram predicto P.'

From which it would appear that Henry III, while on a visit to Wenlock Priory, made a gift of six oak trees, from his forest of Shirlet, to the sacristan of the priory, to build a horlogium, which can be translated as a 'clocktower'. However, whether for a mechanical clock or a water clock, is a question which we are left to conjecture.

It is not until the 16th century that clockmakers in this country begin to appear by name in records, and then only as blacksmiths, engaged in making the great iron church clocks. The large iron turret clock found in so many church towers, being easier to make, and constructed by techniques already known to many blacksmiths, was generally supplied by such a craftsman and repairs were carried out by local men.

The churches of St. Alkmund's and St. Chad's, Shrewsbury, possessed clocks as early as 1436, as an entry in the Bailiff's accounts shows: 'Will. Dawe custodi communis horicudii ecclesie S'c'i Cedde pro termino natalis D'ni, 2s. 6d.'. Also: 'Ric'o. Mascote, custodi horicudii ecclesie S'c'i. Alkmundi pro salario suo pro termino natalis, 2s. 6d.'. One of which was no doubt the 'Shrewsbury Clock' mentioned by Falstaff in Shakespeare's *Henry IV*. The church of St. Lawrence, Ludlow, had a clock in 1469, when eight pence was paid to mend it; sadly it does not mention who made the repair, but it would be no doubt a local blacksmith. Richard Crosse made a wheel for the chimes of this clock in 1565, when he received sixteen pence for his trouble.

A superstitious account of what was probably a storm of wind in 1533, mentions the clock in the tower of St.Alkmund's, Shrewsbury:

> This yere, 1533, upon Twelfe day, in Shrewsbury, the Dyvyll appeared in St. Alkmund's churche there, when the preest was at High Masse, with great tempeste and darknesse, so that as he passed through the churche he mountyd up the steeple of the said church tering the wyers of the said clocke, and put the print of his clawers upon the fourth bell, and took one of the pinnacles away with him, and for the tyme stayed all the bells in the churches within the sayde towne, that they could neither toll nor ring.*

Salopian Telegraph, 1 Jan. 1842

One line of thought is that the early clocks were made through the collaboration of a scientifically-minded monk with a skilled blacksmith. In this county we have mention of such a monk in the person of William Corvehill of Wenlock Priory, whose death in 1546 caused a record to be made of his accomplishments:

> He was well skilled in Geometry, not by speculation but by experience, could make Organs, Clock and Chines. In kerving in Masonry and Silk weaving and painting, & coud make all instrum^{ts} of Musick & was a very patient & Gud Man borne in this Borowe. All this county had a great loss of S^r Wm for he was a good Bellfounder & maker of the frames.*

One can picture such a genius collaborating with a local blacksmith of more than usual intelligence, to construct clocks for local churches. One wonders if he was responsible for the clock installed in the new Guild Hall, at Much Wenlock in 1540, for which the sum of one pound was paid.

As the need for clocks became more widespread it became necessary for the clockmakers to move about the country, possibly moving on regular circuits, making new clocks, and repairing the old. No doubt this accounts for the entry in the Ludlow parish accounts in 1572: 'Payd to a straunger, for mendinge the clocks at M^r baylyeffes apoyntment ijs. viijd'. Poor man, even the clerk could not remember his name. Often when a parish was reached where the clock was to be made or repaired, the clockmaker would make arrangements to use the local blacksmith's forge. This we find to be the case at Prees, when Richard Horton and his son paid a visit in 1671. The churchwarden's accounts note that £2 was

> Paide William Pye ffor the us of his Shop for Richd Horton for Iron and Coles towards repairing of the Clocke and Chimes.

The churchwardens also paid for local help:

> paid to John Sympson for helping all the tyme the clock was mendinge   0. 4. 0.

and again

> paid for carrying water and clay for the Clockmaker   0. 0. 8.

Later, in 1690, William Pye the local blacksmith mends the chimes himself.

*Bodl. Library. MSS. Gough, Salop. 15.

Introduction

It is a pity that the name of the maker of the fine clock which was put up in the Guildhall at Shrewsbury, in 1592, has not been recorded. Depicting the changes of the moon, it must have been of great wonder to the local inhabitants.

> This yeare and about the ennde of August there was made by the baylyffs of Salop a clocke within the guylde hall there wth a diall wthn the hall and two dyalls wthout the hall the one to serve the highe streete market and passars by and the inhabytants there and the oder towards the corne marckett in lycke man', the wch two dials do not only noate how the howres of the daye passethe but also therein the picture of the moone how it dothe increase and decrease verey artyficiall and comodius to the beholders.*

At that time (1592), Adam Brodshawe, blacksmith of Shrewsbury, was a clockmaker of some local repute; in 1594 he made a church clock for St. Oswald's, Owestry, for which he was paid £10; he further agreed to keep it in order for the rest of his natural life (see the indenture under his name in the Biographical List). The public clocks of Oswestry were a source of amusement at one time, the church clock and the Bailey clock never agreeing. Shirley Brooks, journalist and novelist (1816-1874), who was once a resident of Oswestry, explains the discrepancy in his clever novel *The Gordian Knot*.

> We had two public Clocks—one the Church clock, and the other a clock in the Town Hall. The consecrated works were confided to one clockmaker, and the secular wheels to another. Less than five minutes walk took you from one building to the other, but there was always more than five minutes difference between the clocks. The reason was that the two artificers hated each other, and it was merely to annoy the poor old man that he was sometimes asked to explain the discrepancy, but the irreligious horologist not being appointed for life, was obliged to be at once more civil and more scientific, and boldly declared that he kept his clock five minutes a-head of the Church to allow for the difference of longtitude.

Towards the end of the 18th century local directories came into being, and from them we are able to extract a great many fascinating facts about our clock and watchmakers. Now we are able to pinpoint in which streets of a town their businesses were situated, and whose place of business could be more quaint than that of John Massey of Shrewsbury, a watchmaker, who in 1816 plied his craft in the old tower, still standing on Town Walls.

Early Chronicles of Shrewsbury, 1372-1603. Transcribed by Rev. W. A. Leighton, pub. 1880.

It is amusing to see the many and varied sidelines which many of these craftsmen had to take up in order to scrape a decent living for themselves and their families. Joseph Churton of Prees also worked as the landlord of the *New* inn, while William Deakin of Dawley is mentioned as a beer retailer in 1856. Thomas Glaze of Bridgnorth also worked as a gunsmith; Henry Charles Bond took on the respectable sideline of Registrar of Marriages at Church Stretton. An unusual combination was that of James Molesworth Pearson of Bridgnorth who mixed his watchmaking activities with that of being a 'Gilder', and a 'Dentist'. Of course the watch and clockmaker who really stands out from his fellows in his versatility is Robert Webster of Shrewsbury. A fine second generation craftsman he excelled at watch and clockmaking at his premises in Mardol. He expanded his father's business into quite a large manufactory, making large turret clocks for many Shropshire churches and public buildings. He was for a period in partnership with Hazeldine the ironfounder, and was in business as an ironmonger. He was an inventive genius of no mean ability, and is reputed to have made a clock which required winding only once a year. On 12 February 1791 he had the honour of presenting to Queen Caroline, at Windsor Castle, three very elegant spinning wheels, all of his own manufacture—a footwheel, a table-wheel, and a girdle-wheel. The first two wheels were his own inventions, the mechanism and motion was so finely balanced that not the least sound or noise could be heard when they were in use. The Queen most graciously accepted and highly approved of the gift.

Encouraged by this royal patronage, Robert Webster went on to patent in the following year a washing-machine of his own invention. In the words of the specification of 1792, 'A mill or Machine on an entire new construction, for the purpose of washing and cleansing every article which has hitherto been washed by hand, in a much more expedious and less expensive manner than by any machine or method hitherto discovered'. Sad to relate this 18th-century forerunner of the electrically-run washing-machine so commonly in use today did not catch on. The only known example perished in a fire. Not to be daunted Robert Webster went

Introduction

on experimenting with his ideas. In 1812 he was once more at the Patent Office, his mind still wrapped up with washing-day. He then patented 'A Combined Portable Mangle on an entirely new construction for the purpose of mangling every article which has hitherto been mangled in a much more expedious manner than by any other machine or method hitherto discovered'. Sound though his inventions were, it would seem the public of his day were not quite ready for them; his new-found gadgets did not appeal. In 1817 he handed over his watchmaking business to his son, John Baddeley Webster, but with no idea of retiring. At the age of 62 he was all set to open a new business as a manufacturer of brushes, both wholesale and retail. This remarkable man lived into his 78th year, and died in 1832.

These horological craftsmen were not above breaking the law on occasion, and now and then are to be found suffering the penalty of so doing. John Capper who repaired the clock of the Abbey Church at Shrewsbury in 1578, was hung for treason on 24 March 1581. John Middleton, watchmaker of Shrewsbury was one of a large body of the inhabitants of the town who are named on a writ issued by Oliver Cromwell, in 1656, ordering these persons to be apprehended and taken before the Justices of the Peace, 'for certain contempts & other offences'. Another watchmaker of Shrewsbury, Ellis Bradshaw, actually died while a prisoner at the local gaol in 1707. Of a less serious nature perhaps is the following advertisement in the *Salopian Journal* of April 1802:

> Whereas Samuel Pedley Clock Maker is under an agreement to Robert Webster of this Town for near Two years to come, notwithstanding which he scarcely works one-third of his Time, and that without any just cause but the Effect of Idleness and Drunkenness. This is therefore to warn all Persons whomsoever against harbouring or employing him after this notice, as they will be dealt with as the law directs.

And, of course, the other side of the coin, when the craftsmen themselves were the victims: Rebecca Kelvey, watchmaker of Mardol, Shrewsbury, whose husband attacked her viciously; he was taken into custody (County gaol) where he committed suicide in his cell, by hanging (March 1850).

Watchmakers and clockmakers are to be found in public office. Francis Campbell became mayor of Oswestry in 1836; at Shrewsbury, Samuel Harley was mayor in 1784; to be followed by his son, William Harley in 1814. John Green, watchmaker, served for a year (1688) as constable, keeping the peace inside the Castle Ward Within the Walls, at Shrewsbury, while many of these tradesmen dutifully did their stint as churchwardens of their parish churches.

Most of the clock and watchmaking businesses only lasted for the working life of the tradesmen, but here and there we find two, three, and sometimes four generations of a family following the craft of making clocks and watches, and at times even inter-marrying with other families engaged in the same trade.

The Webster family of Shrewsbury, established in 1740, were clockmakers for four generations, the last craftsman dying in 1871. The Harleys, also of Shrewsbury, provided three generations of clockmakers during the 18th century. Robert Edwards, watchmaker of Oswestry trained four of his sons to follow him in his business, while William Davis brought up three sons as clockmakers at Shifnal. After their apprenticeships were finished they spread about the county and established their own businesses at Shrewsbury, Wellington, and Shifnal. The name of Evans was a common sight above the 19th-century watchmakers' shops at Shrewsbury: William Evans, clockmaker of Gullet Shut was followed by three sons, all in the same trade; James Evans in the Cornmarket, son of a Dissenting Minister of Oswestry, trained his three sons in his own craft; while yet another Evans, Richard by name, passed on his business to his son.

But, of course, when it comes to speaking of a family business handed down from son to son, generation to generation, there is no one who can even pretend to compete with the Joyce family of Whitchurch in the clockmaking sphere—eight generations, from which came 25 clock and watchmakers, spanning the reigns of 14 monarchs.

In *Bye-Gones* of 1903, there is a vague reference to a club at Shrewsbury, with the unusual name of 'The Watch and Clock Club'. The reference is from an old diary, author unknown, in which he states:

Introduction

> Oct. 6th (1810). Attended supper of the Watch and Clock club at Atcham.

He buys a new watch from Henry Rowley, watchmaker of Shrewsbury, on 1 December 1810; Rowley appears by the next entry to be a member of the club.

> Feb. 3 (1811). Went with Roberts, Blacksmith, and George Roberts to the Talbot, where Rowley the watchmaker was to have met us respecting the Clock and Watch Club, but he did not come; so we left him 4 jugs of ale to pay for.

Poor Rowley, a maker of watches, but a poor timekeeper.

It is of some interest to compare the will and inventory of George Birchall, watchmaker of Shrewsbury, dated 1738, with that of Edmund Bullock, clockmaker of Ellesmere, some four years earlier. The first is that of a watchmaker of the county town; the second that of a more countrified clockmaker, who includes in his inventory cattle and grain:

> In the Name of God Amen. I George Birtshall of the Town of Shrewsbury in the County of Salop Watchmaker Do Give Devize and bequeath all my Goods Chattells and personal Estate of what kind or Nature Soever I shall dye pofsessed of or intituled unto, unto my Executor hereinafter named In Trust neverthelefs to be Sold and equally divided between my three Daughters Margaret, Hannah and Elizabeth Share and Share alike and the Survivors and Survivor of them living to Attain the Age of one and Twenty years or to be Married which shall first happen And the Interest in the meantime for their Education and Maintenance And do hereby Nominate and Appoint Richard Mall of Middle in the County of Salop Executor of this which I do Hereby declare to be my last Will and Testament In Witnefs whereof I have hereunto Set my hand and Seal the Sixteenth day of February in the Year of our Lord One Thousand Seven hundred and Thirty Six. Signed Sealed published and declared by the abovenamed Testator as his laft Will and Testament in the prefence of us who have Subfcribed our Names as Witnefses thereto in the said Testators prefence.   Geo: Birchall   Will: Prichard   Robert Corbett   Thomas Morris.
>
> This is my Codicil to my Will
> I Leave to my wife Elisabeth Birchall (as I Left her nothing in my Will) the yearly Summ of three pounds to be paid her Quarterly as Long as she shall live by my Executor in my will nam'd as witnef my hand and Seal This 18th July 1738. Sign's and Seal'd by the said Testator as a Codicill to his Last will in the presence of us who Sub- scribed our names as witnefses thereto.   Will. Prichard.   Tho: Burgefs Junr.   Tho: Bridges.   George Birchall's mark.

With the will is attached an inventory of all his goods, giving a clear picture of the contents of both his shop and his house.

A true and perfect Inventory of the Goods and Chattels of George Birchall Watchmaker taken and Appraifed the Thirty first day of July Anno Dni 1738.

	£	s.	d.
A Spring Clock	10	0	0
one eight day Clock with a fineard case	5	10	0
one eight day Clock with a black case	3	0	0
two Eight day Clocks without Cafes	6	0	0
One Eight day Clock Archt Diall plate	3	10	0
one thirty hour Clock with the day of the Month	2	0	0
two time peices			
one time peice in the Shop			
one Clock Engine)			
one Barrell Engine)	1	0	0
one Watch Engine			
one Watch ffufee Engine			
one Ballance Wheel Engine			
one large pair of Turn Bench (Lathes)	0	6	0
one Small pair of Turn Benches	0	3	0
one Beam Compafs	0	1	0
two pair of Turn Benches	0	6	0
Three hand Vices	0	2	6
one large Smiths Vice			
ffour board Vices	0	16	0
one Small Smith Vice			
two Saw fframes	0	2	0
one peircing Saw fframe			
two planishing Tools..	0	1	0
two pair of Sliding Tongs	0	0	10
one pair of Nipars	0	0	6
Three pair of Callipers	0	0	9
a Tool for the breath of pelletts	0	0	6
two Tools for tempering pendulam Springs	0	1	0
one Small pair of Turnbenches	0	1	6
one watch kepeter	0	12	0
an Engine—a Small Jack both imperfect Six Bells	0	4	6
one twelve Inch plate & two Circles, one ten Inch plate & Circle & other Cast brafs weighed 24^{li}.3 at 15^d p £.	1	10	11
one boiling pot Copper 6oz	0	0	6
uncast brafs 1^{li}. 1^{qr}	0	1	8
a Silver Watch Glafs in Cock n°. 283	3	10	0½

continued—

	£	s.	d.
A Silver Watch Dent Nº 306	2	5	0
A Bathmettle Watch Nº 222	4	0	0
A Silver Watch Nº 277 Vaughan ..	3	10	0
A Silver watch Nº 308 GB ..	3	10	0
one Dent Nº 294 finished ..	3	10	0
one Birchall Nº 302 Mr. Gorsuch	3	10	0
one Single Cafe Nº 298	3	10	0
Nº 310 not Gilt	2	15	0
Not finished Nº 301 ..	3	0	0
one Gold Watch Nº 297	11	0	0
one Gold Watch Nº 160	8	0	0
Jones 305 not finished	2	15	0
one 3 1 not finished ..	2	10	0
A Movement box & Cafe not graved	2	0	0
A Watch Birchall Salop on the Diall) plate a box but no cafe)	2	0	0
A Movement & box ..	1	7	6
four frames motioned			
two pairs of old boxes & Cafes			
an old String Watch & box			

In the paper Room upstairs.

	£	s.	d.
A blew Curtain Bed hath two feather beds) a bolster only a Single furniture)	2	0	0
one large looking Glafs	0	10	0
a prefs Cupboard	0	2	6
a Chest of Drawers	0	10	6
a Chest to hold Linnen	0	3	0
a Cloathes Prefs	0	2	6
a Clofe Stool	0	2	6
An Ordinary Trunk four boxes three Chairs	0	2	6

In the Garrett.

Three feather beds & 2 bolsters at 7^d p £

In the paper Room up one pair of Stairs.

	£	s.	d.
one Chest of Drawers	0	15	0
one Rusty Jack in the Clofet	0	15	0
a Glass bofett ..	0	10	0

In the Green Room.

	£	s.	d.
one Green bed and furniture	2	0	0
one Ovall Table	0	4	0
a pair of Tongs fire shovell & fender	0	1	6

continued—

	£	s.	d.
an Ordinary Square Table	0	1	0
five Ordinary Afa (Ash?) Chairs	0	2	6

In the parlour.

	£	s.	d.
one Couch with a Squab	0	4	0
one Ovall Table	0	5	0
five Cane Chairs	0	8	0
a brafs Jack	0	15	0
2 Cuboards	0	2	0
picture frames			
a fender fire shovell & tongs	0	3	0
one bedsted & a Clofe Stool	0	3	0
three large Silver Spoons	0	18	0
a pair of Silver Tongs three Silver Tea Spoons ..	0	10	6
a Small hand bell	0	1	0

In the Kitchen.

	£	s.	d.
a pair of Bellows	0	1	6
A warming pan	0	7	6
a Beauvois	1	10	0
two Iron plates			
a gridiron	0	0	8
a Coffee pott			
a box Iron heater & Stand	0	1	6
two flatt Irons	0	2	0
1 flower box 2 pudding pans 1 Cullender)			
1 Lanthorn A Grater 2 Small brafs Candlefticks) ..	0	6	0
1 large Square Do.)			
4 more Common Do.			
12 pewter dishes			
18 plates & a Salver			
2 pewter porringers			
a Candlebox			
a pewter Mustard pott			
a perper box			
2 tin Tunning Dishes..	0	1	0
A Tin Oven			
A Tea kettle	0	1	6
a brafs Mortar	0	0	6
a Tofsing pan	0	2	0
2 brafs Kettles..	0	8	0
1 Mafhing kettle	0	8	0
1 brafs Spoon	0	0	3
A Chopping knife	0	0	6
a frying pan	0	0	6

continued—

Introduction

	£	s.	d.
2 Sauce pan	0	0	6
a flefh fork an Iron Egg Spoon	0	0	6
2 Spitts	0	3	0
4 knives & 6 forks	0	2	0
a large Chafting Difh..;	0	2	0

In the Buttery.

a Cupboard	0	1	6
a board tinn'd to put before meat	0	1	6
2 leather Chairs	0	2	0
3 Wainfcott Chairs	0	2	6
a pewter Chamber pott			
a Saltbox	0	0	4
a led Scure	0	0	2
an Ordinary Dale (Deal) Tea Table	0	0	6
2 hangings in the parlour	0	0	9
one barrell & 2 half barrells	0	7	6

In the Shop.

one broken Ovall Table	0	0	9
1 Square table	0	2	0
a Chest	0	1	6
a Dale box	0	0	6

In Mr. Powells Custody.

6 pewter Dishes at 8^d p £			
12 pewter plates at 10^s p Doz^n	0	10	0
a Square oblong table in the Kitchen	0	2	6
a pitgrate & key at 4^d p p^d			
a Spitt & hanging Racks	0	3	6
a brafs Jack & Weights	0	15	0
a Skreen in the Kitchen	0	2	0
a Camlet bed & Curtains & bedSteads	2	10	0
a feather bed & bolster at 7^d p pd			
Jericho's picture			
6 black pictures	0	3	0
Window Curtains & hangings for the room)			
paper hangings in the Clofet)			
a Garrett bedftead & Matt	0	5	6
an old Stool in the Cellar	0	0	6
a Drefsing table in the room	0	2	6
Book Debts	5	0	0
Cash			
wearing Aparell	1	0	0
Lumber			

The above inventory is not signed by the appraiser, nor is it totalled up, which is unusual.

Another Inventory of interest is that of Edmund Bullock, a fine clockmaker, who died in the year 1734, at Ellesmere.

A True and perfect Inventory of all and singular the Goods Chattles & Creditts whereof Edmund Bullock late of Ellesmere in the County of Salop dyed pofsefsed. Taken and appraised the (blank) day of (blank) in the year of our Lord 1734 by the persons whose names are underwritten.

	£	s.	d.
Five clock cases	5	17	6
Six working vices	2	0	0
Two Anvils	2	18	0
Three wheels Leather	2	2	0
Two pairs of large Bellows	1	6	0
Four Engines and Engine files	1	17	6
Two turnbenches	1	1	0
Old Brass	2	18	6
Flasks and Boards	0	12	0
Brafs Heads Paterns	0	10	6
Brafs & Led & Wooden paterns	3	2	0
More Working tools	5	6	9
Brafs Work more	0	12	0
five pair of Scales	1	8	6
Brafs Weights	0	2	4
A Cole Trough & Sand Trough	0	4	6
files	0	3	0
two Grinding Stones	0	7	0
A large Leather & other tools b'longing to it	0	15	0
A anvil Block & polishing Boards	0	1	0
Clock Hands	0	3	0
Old Silver & Gold	0	9	0
Clock Line & Silvering	0	1	0
New Files	1	6	6
three new Jacks	3	10	0
Swifles	0	2	9
Watch Springs Keys & Glajses & Wax	0	6	10
two parcells of Wyer	0	2	0
bowell string	0	4	0
brafs patterns	0	2	0
Wood frifses	0	1	6
A watch	3	3	0
new Clocks	18	0	0
two old watch parts	0	7	6
files and other odd things	0	2	6
A weather glafs Varnish and Aquafortis	0	8	0

continued—

Introduction

	£	s.	d.
two Sundials ..	0	2	0
A peice of Box tree and Oil	0	2	0
Crufibles and Other things ..	0	2	6
Iron ware and a small parcell of tools ..	0	6	0
A Bell ..	0	1	6
In the Garret, two pairs of Bedsteeds and one bed ..	1	0	0
two Sashes ..	0	13	0
boards and other odd things	0	6	0
In another Garret, one bed..	0	6	0
a parcell of old Iron ..	0	10	0
For Casks and other things..	0	6	0
for what remains in one Garret ..	1	5	0
for what things remains in another garret	0	17	6
In the Closet & Sceutore (secretaire) ..	1	0	0
A hanging prefs and Bedsteeds ..	0	12	6
A parcell of Cheese ..	4	10	0
In the large room, Two Beds ..	5	5	0
a Chefter Draws & Drefsing Table	1	2	0
half a Dozen of Chairs	0	5	0
A Cheft Clofe Stool & a Straw Chair ..	0	12	6
In a Little Room a bed & Bedsteeds & a Desk	1	12	6
Books ..	1	1	0
Linnen ..	4	10	0
What Remains in a Buttery	0	3	0
What Remains in the Sculery	0	17	0
In the Kitching four Iron pots & a boiler	0	10	0
A frying pan & other things	0	3	0
A Dishboard ..	0	12	0
Pewter ..	2	12	0
Brafs belonging to the kitching ..	0	16	0
Tables & a Screen ..	0	9	0
three Spinning Wheels	0	6	0
for a kitching grate & other Iron things belonging to ye fire place	1	8	6
Other Od things	0	5	6
In the parler a Writing Desk & a Corner Cobert	2	0	0
A Looking Glafs & a parcel of pictures..	0	18	0
A Chair & a table ..	0	3	0
In the Celler Horfes & Barrels & Tubs ..	0	19	0
In the Brewhoufe a furnace	1	0	0
Wooden Vessels ..	0	19	0
In the Stable two Sadles & Bridles and pillins	0	18	0
Boards ..	0	5	0
Hay ..	4	0	0

continued—

		£	s.	d.
two Stone Troughs	.. ..	0	4	0
A Counter & Boxes	.. ..	0	18	0
Cattle	.. ..	17	5	0
Oates on the field	.. ..	2	15	0
Manure for Land	.. ..	0	12	0
Coles	.. ..	1	10	0
Plate	.. ..	3	17	6
Wareing Apparell	.. ..	4	0	0
	To.^{ll}	138	8	2

William Bullock⎫
Thomas Edwards⎭ Appraisors.

This Inventory was Exhibited by Sarah Bullock Wid: the Admin^{ex} upon oath at the Court Baron held for the Mannor of Ellesmere the 16th day of August 1734 for a true and perfect Inventory upon protesting nevertheless to add thereto if any further afsets shou'd come to her hand.

The obligation made between the Steward of the Manor of Ellesmere & Sarah Bullock, the widow, is witnessed by two other clockmakers—John Joyce (1718-1787) & Thomas Studley both of Ellesmere.

Often agreements were drawn up between the churchwardens of a parish and a local clockmaker, whereby he was to attend the church clock for a specified number of years. These agreements are quite detailed and leave nothing to chance, as will be seen by the following example:

> Articles of Agreement made the eleventh day of June in the Year of our Lord One thousand Seven hundred and Fifty Nine. Between Richard Stephens of Bridgnorth in the County of Salop, Clockmaker of the one Part. and John Lowe and John Bach Churchwardens and Thomas Martin and William Crow Overseers of the Parish of Alverley in the said County of Salop of the other Part. Imprimis. It is agreed by and between the said Partys to these Presents that the said Richard Stephens in consideration that the said John Lowe, John Bach, Thomas Martin and William Crow on the behalf of themselves and their Successors have agreed to pay yearly to the said Richard Stephens during his natural Life upon every Thirtyeth Day of May in cash and every year the sum of five shillings of Lawful Money of Great Britain and the first payment to begin and be made on the Thirtyeth day of May next Ensuing the Date hereof he the said Richard Stephens hath agreed and undertaken from time to time during his Life to repair

amend and keep in good order and repair the Clock belonging to the Parish Church of Alverley aforesaid commonly called the Church Clock. Except as hereinafter Excepted he the said Richard Stephens in consideration of the Agreements aforesaid on his Part to be performed Doth Covenant Promise and Agree to and with the said John Lowe, John Bach. Thomas Martin and William Crow and their succefsors that the said Richard Stephens Shall and Will from time to time during his Natural Life at his own Cost and Charges as oft as need on Occasions shall require repair and amend and in good order keep the said Clock during the Time and Term aforesaid pursuant and according to his said Agreement and Indempnify them the said John Lowe, John Bach, Thomas Martin and William Crow and their succefsors and other the Inhabitants of the said Parish for the time being of and from the Same Provided they so long continue to pay the said Five Shillings a Year to the said Richard Stephens according to the said Agreement/Save and Except such Damage as Shall or may be done to the said Clock by the Ringing of the Bells. And the said John Lowe, John Bach, Thomas Martin and William Crow in consideration of the said Agreement on their Parts to be performed do as far as in them lies and they lawfully may for them and their succefsors covenant promise and agree to and with the said Richard Stephens his Executors and Administrators that they the said John Lowe, John Bach, Thomas Martin and William Crow and their Succefsors shall and Will Yearly and every Year during the Natural Life of the said Richard Stephens well and truly pay or cause to be paid unto the said Richard Stephens the Sum of Five Shillings upon the Thirtyeth Day of May in cash and every year and in manner aforesaid according to the true intent and meaning of these Presents.

In Witnefs whereof the Partys to these Presents their Hands and Seals interchangeably hereunto have Set the Day and Year first above written.

Sealed and Delivered by the above named
Richard Stephens John Lowe John Bach and
Thomas Martin in the Presence of us
    Thomas Gitton
    John Lindon

Richard Stephens
John Lowe
John Bach
Thomas martin
William Crow

Sealed and delivered by the above Named

William Crow in the Presence of us— The Marke of John Jordan
                                     The Marke of James Power.

THE COMPANY OF SMITHS, SHREWSBURY

Founded in the year 1621.

This Shrewsbury fraternity of tradesmen had become known by 1656 as 'The Company of Smiths, Farriers, Armourers, Cutlers, Furbishers, Spurriers, Sheathmakers, Sheathgrinders, Braziers & Clockmakers'.

The Minute and Apprenticeship book, dated 1622-1764, does not include clockmakers (as such) until the opening of the 18th century.

> William savage puts himselfe An apprentice unto his father Richard Savage Clockemaker for seaven years his time to Begine the 23rd of June 1700
>
> Thomas Gosage came in free that yeare as a Clock Maker. 1701

> February 24th 1702/3.
>
> Richard Savadge hath putt himself an aprentice to his father Richard Savage clockmaker for the term of Seven years his time beginning the day above written

> Thomas Savage Son of Rich. Savage Clockmaker putt him Selfe an A prentis to His father for Seven years is time Begining 30 of December 1706

> Joshua Johnson hath put himselfe aprentise to Richard Savage Clockmaker & his time to begine the 16 of November 1709

> William Harley hath Put him Self An Apprentice to Thomas Gorsuch Clockmaker in Shrewsbury for Seaven Yeares his time begins November the 11th 1717   Rec.^{d} 0. 2. 6.

> Samuel Brodhurst prentise to Tho: Gorsuch Clockmaker for seven years his time beginning ye 2nd Day of may 1718.   0-2-6

> William Harris son of Richard Harris hath with the Consent of his father put himself aprentis to Thos. Harper Clockmaker. His time to begin the 2nd day of Feby 1723/4.

Introduction 19

May 25th 1725. Thomas Unett son of Anne Unett of wolverhampton in the County of Staffordshire widdow hath put himself an apprentice to Thomas Gorsuch of the Town of Shrewsbury a Clockmaker beging the 25th day of may according to the date of his Indenture.

John moody son of Sarah moody widow hath put him selfe an apprentis To Thomas Nash Clockmaker his time to begin the 8th of June Acording to ye date of his Indentures 1725.

Benjamin Tipton hath putt himself an apprentres To Tho: Nash Clockmaker for seven years is Time begins ye first day of August according to the date of is Indenteurs.   2s. 6d.
May ye 5. 1729.

Tho. Traunter son of Robert Traunter of Welshpoole In Moungomry hath putt him self an aprentice to George Birchall Clock maker for seven years Is time begins according to deate of Is Indenture Recd. 0-2-6.

Willm Davis sett himselfe aprentise to Willm Harley ClockMaker for ye terme of seaven years bearing date as his Indentures. (*c.* 1731).

Charles Barratt hath put him Self an aprentice to Jas. Evans Watch & Clockmaker he giving at ye date of his Indenture.
Recev.^{d} for the Use of the Company 5s.   (*c.* 1747)

1748. Richard Harper son of Thos. Harper hath putt him Self an Aprentis To his ffather for The Term of Seven Years.
Recev^d. for the Use of the Company 5s.

Mathew Holland hath Put him an Aprentess to Richard Wood for the Space of Seven years according to the Date of his Indenture.
Recev'd for the Use of the Cumpany 5s.   (*c.* 1748)

Jas. Bearley hath put him Self an Apprentice to Jas. Evans Clock & Watchmaker. May 26. 1748. Rec^d for the Use of the Cumpany 5s.

John Dickin hath put himself Aprentis To William Harley Watchmaker for the term of Seven Years Bearing the Date of his indenter Rece^d. for the Use of the Company 5s.   (*c.* 1749)

Thomas Harper, watchmaker, appears as one of the two wardens of the company in 1748.

Of the above apprentices both Thomas and Richard Savage died while still in their apprenticeships. William Harley prospered as a watchmaker in Shrewsbury. John Moody set up in business at Oswestry. Benjamin Tipton appears at Ludlow as a clockmaker. William Harris, Richard Harper, Thomas Traunter, Charles Barret all set up in trade at Shrewsbury, while the others disappear into obscurity, their future unknown.

THE COMPANY OF HAMMERMEN, LUDLOW

At some time prior to the reign of Richard I, the blacksmiths of Ludlow, in company with other allied trades, combined together to form a trading fraternity or guild for their mutual protection and benefit.

They became in time a body corporate, known as 'The Antiente Company of Smiths and Others Commonly Called Hammermen'.

Governed by the 'Six Men', they elected two stewards every two years to look after their interests. Their meetings were held in the south aisle of St. Lawrence's church. Such fraternities enjoyed great privileges, and tradesmen who were not freemasters of the guild were debarred from trading within the town of Ludlow.

Within the books of this ancient body are to be found both the enrolments of apprentices, and the admittances of freemasters, including from the opening of the 18th century those of Clock and Watchmakers, which have been extracted as below.

> Att a meeting the ninth day of Sept[br] one thousand Seven hund[d] & Ten Agreed therein by ye six men and the rest of the Company then p'sent That William Brodhurst is admitted A ffree master to the Trade of A Clockmak[r] & Watchm[ker] hee handing unto the Steward the Sum of five pounds.　　　Will. Brodhurst.

22nd of March 1711.
> At a Meeteing then held for the Company ffees of Smyths & Others It was agreed by the Six Men and the Rest of the Company then p'sent that Thomas Vernon be Admitted a ffreemaster to a Clocke & watchmaker a Member of ye sayd Company he paying to the present Steward Crow three pounds for ye use of the company (fforty shillings being already payd to George Cooke the former Steward towards his fine for his ffreedom this being five pounds.
> 　　　　　　　　　　　　　　　　　　　　　Tho. Vernon.

Introduction

25. Junij 1717.

Memorand that then Edward Stead was Enrolled haveing Set himselfe an Apprentice to Thomas Vernon to ye trade of a Clock & watchmaker for ye terme of Seven years by Ind: dated 7th day of octo 1715.   Edward Stead.

24 Julij 1717.

Then Ric^d Felton was Enrolled haveing set himself apprentice to Thome Vernon to ye trade of Goldsmith Clock & Watchmaker for ye terme of seven years by Ind: bearing date ye first day of may 1711.   Richard felton.

17. Junij 1718.

Then Rich^d ffelton was admitted a ffreem^n to ye trade of a Goldsmith Clock & Watchmaker having served an apprenticeship to Mr Thomes Vernon one of ye ffreem^n of this Company paying to Mr Steward James Wilcox Twenty shillings to ye use of ye Company.   Richard ffelton.

James Noke sonne of Wm Noke was Inrolled an apprentice to Thomas Vernon by the Trade of Clock, Watchmaker & Goldsmith by Ind: bearing date 11. Nov: 1718.

29. Maij 1722.

Then Wm Percival was Admitted a ffreem^n to ye Trade of a Clockmaker having been an apprentice & working at ye trade twelve years or upwards paying as a fforeigner to ye Steward Mr. Jon Hattam ye sume of Tenn pounds to ye use of ye Company
     per me Will Percival.

26 Decem. 1722.

Then Edward Stead was Admitted a ffreem^n to ye trade of a Clock & Watchmaker paying 20s. to Mr. Steward Hattam for his ffreedome.   Ed. Stead.

11th. Aprill 1726.

Then James Noake was admitted a Freeman to the Trade of Clockmaker and Watchmaker haveing Served a Legall Apprenticeship to Mr. Thomas Vernon of the same Town by Indenture bearing date the Eleventh day of November One Thousand and Seven hundred & Eighteen.   James Noke.

1. October 1737.

Then the Company met being Quarter Day in the usuall place in the Church and made George Payne who Served a Regular Apprenticeshipp to Mr. Thomas Vernon a Freemaster of This Company to a Clock and Watch-Maker He payeing a Fine of Twenty Shillings to the Use of the Company to Mr. Steward Prodgers.   Geo. Payne.

5th June 1771.

Then William Langford (Son of Samuel Langford who was a Freeman of this Town) was admitted a freemaster of this Company to the Trade of a Clock and Watchmaker he paying a ffine of one pound which he accordingly did to Mr. Steward Hudson to the use of the company. Wm Langford.

5th May 1778.

Then William Camell was admitted a Freemaster of this Company to the Trade of a Clock and Watchmaker he foregoing down a ffine of three pounds (two pounds being abated him out of his ffine of five pounds) he being a Forreigner which he accordingly did.

William Camell.

A list of admittances dated 1790–1838 include the names of:

William Herbert, watchmaker, admitted 26th December 1820.

William Payne, watchmaker, admitted 26th December 1820.

Edmund Lechmere Charlton, watchmaker, admitted 5th July 1830.

Stewards' appointments are listed as follows:

Benjamin Tipton, Clock and Watchmaker in 1747.

George Payne, Clock and Watchmaker in 1749.

THE WORSHIPFUL CLOCKMAKERS' COMPANY, LONDON

Incorporated by Charter 1631

The register of apprentices of this London guild or company lists the names of five Shropshire boys who were apprenticed to London craftsmen.

George Guest. 25 Mch. 1762, son of Michael Guest of Bridgnorth, Salop, sadler, decd. apprenticed to Boys Err May. £31. 10s. for 7 years.

Robert Jeffryes, 25. June, 1772, son of Robert Jeffryes, of Shrewsbury, shoemaker, decd. apprenticed to Joseph Robinson, Hosier lane. £10. for 7 years.

William Langford, 2. June, 1761, son of Samuel Langford, of Ludlow, apothecary. apprenticed to Pointer Baker. £10. 10s. for 7 years. Freeman. 22. Oct: 1770.

William Powell. 7. Nov. 1803, son of William Powell, of Shrewsbury, labourer, decd. apprenticed to Joseph York Hatton, of Tooley Street. One Penny.

Thomas Vernon. 20. May, 1701, to Robert Halsted. for 7 years. Freeman. 2. August, 1708.
(Possibly the Thomas Vernon admitted to the Company of Hammermen, Ludlow—22 March, 1711.)

William Langford (above) also became a Freemaster of Ludlow in 1771, when he paid a fine of one pound for the privilege.

* * * * *

No Person shall after the Twenty fourth day of June, 1698, Export or send out of this Kingdom, any outward or inward Box, Case, or Dial-Plate of Gold, Silver, Brass, or other Metal for Clock or Watch without the Movement in or with every such Box, Case, or Dial-Plate, made up fit for use, with the Clock or Watch-Makers name Ingraven thereon:

Nor shall any Persons after the said Twenty fourth day of June make up, or cause to be made up any Clock or Watch without Ingraving or putting their own Name or Place, on every Clock or Watch they shall so make up, on the forfeiting every such Empty Box, Case and Dial-Plate, Clock and Watch not made up and Ingraven as aforesaid, and for every such Offence Twenty Pounds, one Moiety to the King, the other to them that shall Sue for the same, in any of His Majesties Courts of Record.

(Statute. 9 & 10. William III. cap. 28)

It is thanks to the promulgation of the above statute in 1698 that today we are able to recognise the clocks of individual makers, and value them as examples of their workmanship. Certainly a lot of the interest in old clocks would be lost if we did not know the makers, or the towns in which they worked.

BIOGRAPHICAL LIST OF CLOCK AND WATCHMAKERS OF SHROPSHIRE

Abbreviations

BCR	Bridgnorth Corporation Records	LV.M	Liverpool Museum
BG	Bye-Gones (Shropshire)	LWM	Ludlow Primitive Methodist Registers
BG Oswestry	Bye-Gones of Oswestry,	O & B	Owen and Blakeway, *History of Shrewsbury*
BL	Shrewsbury Burgess Rolls		
C	Census Returns	PB	Shrewsbury Poll Books
CC	Company of Clockmakers (Register of Apprentices)	R	Parish Registers
		SC	Shrewsbury Chronicle
		SJ	Salopian Journal
CCA	Chirk Castle Accounts	SM	Salopian Magazine
CBR	Company of Blacksmiths (Shrewsbury) Records	SRO	Salop Record Office
		SBL	Shrewsbury Borough Library
CS	The Churches of Shropshire (Cranage)	SN & CR	Shrewsbury News and Cambrian Reporter
CW	Churchwarden's Accounts		
CHM	Clive House Museum, Shrewsbury	SQS	Shropshire Quarter Sessions Papers
D	Directories	SR.BR	Sessions Roll, Shrewsbury Borough Records
EJ	Eddowes Journal		
EL	Electors Lists	UBD	Universal British Directory
GHR	Guild of Hammermen (Ludlow) Records	VL	Voters' List
		VM	Vestry Minutes

Anderton, William   Shrewsbury, 1871–
'W. Anderton, Practical Watchmaker, & Working Jeweller, 49, High Street. (formerly for 3 years with E. H. Robinson of the Square).' Advert: *SJ,* 19 July 1871.

Addison, John   Bridgnorth, 1841–1851
'John Addison, age 35, watchmaker, Waterloo Terrace. Mary, wife, age 50. Mary dau: age 14, Caroline dau: age 12, James son age 10.' (C. 1841.) John Addison, watchmaker, Waterloo Terrace. 1842–1846. (D). St. Mary St. 1849–1850. (D). High St. 1851. (D).

Ames, Thomas Cleobury Mortimer, 1768-1780
Watchmaker. Listed by Baillie an. 1780. Possibly the Thomas Amis who married Maria Griffiths 1768. (R).

Andrewes, Edmund Shrewsbury, 1646-1650
Watchmaker of St. Chad's parish. Children by Eleanor, his wife: Ann bur. 1646; Edmund bapt. 1647; Eglingtinne bapt. 1649; Ann bapt. 1650. (R).

Aris, Philip Shrewsbury, 1815
Clockmaker of Double Butcher Row. Child by wife, Mary: Margaret Ellen bapt. 1815. (R).

Armstrong, William Ellesmere, 1841
William Armstrong, Watchmaker, aged 20. (C 1841.)

Arkinstall, Francis Market Drayton, 1807-1846
Watchmaker, High St. 1822-1846. (D). 'On the 19th ult. Mr. Arkinstall, clock & watchmaker, of Market Drayton, to Miss Groom, milliner & dress-maker, of that place.' (*SJ*, 2 December 1807.) A watch stolen at Newport Fair in the June of 1829, made by this maker, was identified by him at the County Quarter Sessions as having been sold by him to a Mr. Thorpe (a silver watch with a metal chain, and two steel swivels and a steel seal attached, worth £2).

Francis Arkinstall, Clockmaker, age 60, of High Street. Maria Arkinstall, aged 25. (C. 1841.)

Ashby, John Ludlow, 1851
Watchmaker of Raven Lane 1851. (D). 'John Ashby, Raven Lane, Watchmaker, aged 27, married, born Coventry. Rebecca Susannah Ashby, wife, aged 27, born Coventry. James John Ashby, son, aged 3, born Ludlow.' (C 1851.)

Aspinall, William Shrewsbury, 1623
William Aspinall, of Shrewsbury, watchmaker. Benefactor to Shrewsbury School Library, 1623. (*Salopian Shreds and Patches*, Vol. 1, p. 24.)

Biographical List of Clock and Watchmakers

Aston, Samuel   **Ludlow, 1851-1861**
'Samuel Aston, Watchmaker, Marr: age 56, Lower Broad St. born Frodsham, Cheshire. Mary Aston, wife, aged 47, born Shifnal.' (C. 1851.) Similar entry on Census Return for 1861, residence now Corve St. The 1861 Return spells his name as 'Aster', but the 1861 as 'Aston'.

Baddeley, George   **Newport, 1730-1785**
Clockmaker. Son of George Baddeley, blacksmith, and Sarah his wife, bapt. 1730 at Tong. (R). Married 1754 1/ Mary Webster. Children: Anne 1755, bur. 1760; Mary 1757, bur. 1758; Sarah 1759; Robert 1761. His wife, Mary, died 1763. Married 2/ 1764, Martha Shaw. Children: George 1765; John 1766, bur. 1766, John 1768; Thomas 1769; William 1771, bur. 1772. His second wife, Martha, died 1788. George Baddeley died 14 September 1785. (R.)

Baddeley, John   **Tong and Albrighton, 1727-1804**
Clock and watchmaker. Son of George Baddeley, blacksmith, and Sarah his wife, bapt. 20 September 1727 at Tong. (R). Long case clock, dead beat escapement, about 1750, signed 'J. Baddeley, Tong'. (Britten.) Also a fine Long case clock, 8 day, brass face, inscribed 'J. Baddeley. Tong.' (CHM). Children by Martha, his wife: Martha 1752, married 1776, to Robert Webster, clockmaker of Shifnal (q.v.); and Anne 1756, married 1781 to Thomas Underhill, watchmaker of Albrighton (q.v.). John Baddeley removed from Tong to Albrighton about the year 1766, to premises situated in the already appropriately named 'Clock Mills', an ancient property which had been so called since at least 1623. This maker was responsible for the clock on the tower of Albrighton church. He was buried at Albrighton '30. Jan: 1804 John Baddely, gent. 76. buried'. (R). The Baddeleys were a prolific family, and there appear to have been several John Baddeleys, clockmakers, which make for confusion in ascribing a clock to one particular man. In 1797 a John Baddeley, jun. of Albrighton married Ann Childe (or Childs) at Pattingham, co. Staffs; they had a son, again John, in 1803. There is also recorded in the Pattingham registers the death of a John Baddeley in 1804, and that of an Anne

Baddeley in 1805. A couple of decades later there appears in the local newspaper: 'On the 31st ult. much respected, Mr. John Baddeley, clock and watchmaker, Pattingham' (*SJ*, 8 Jan. 1828.)

Baddeley, Thomas   Albrighton, 1795
Clockmaker. Noted by Baillie.

Baddeley, Thomas   Shifnal,1789
Clockmaker. Listed in the *Universal British Directory*, compiled 1789, published 1790.

Baker, William   Shrewsbury, 1821-1857
Watchmaker, silversmith and cutler, Cornmarket. Succeeded to the business of his uncle, Robert Morris, silversmith and cutler, January 1821. He had previously been with a London watchmaker and jeweller for some 20 years. Watchpaper: 'Wm. Baker, Silversmith, jeweller, Cornmarket. Shrewsbury.' (CMH). Watch movement: 'Wm. Baker. Shrewsbury', inscribed on back 'Emma Eddowes 1829'. (CHM). He appears on the Electoral Lists for 1831 and 1832. William Baker, Shrewsbury. Early 19c. Watch. (Baillie.) His watchpapers generally depicted the Market Hall, which was opposite his shop. 'BAKER.—9th April, aged 68, Mr. William Baker, silversmith, Market Square; one of the oldest & most respected tradesmen in this town; regretted by a large circle of friends.' (*SJ*, 15 April 1857.)

Banister, James   Ellesmere, 1773-1780
Watchmaker of the High Street, Wrexham. Churchwarden at Wrexham, 1765-6. Noted by Baillie at Wrexham an. 1768, and at Ellesmere in 1773. He died in 1780.

Banks, John   Dawley, *c*. 1817-1879
Watchmaker of King Street 1856-1879 (D). The Census Return of 1861 lists him as: 'John Banks, King Street, aged 44, marr: Watchmaker, born Dawley *c*. 1817.
Elizabeth Banks, wife, aged 39, born Painslane, Salop.
Thomas Banks, son, aged 16, born Dawley.
John Banks, son, aged 14, also born Dawley.
Jabez Banks, son, aged 9, born Dawley.'
The Census of 1871 lists two more children: Jessie Banks, dau., aged nine, and Emma Banks, dau., aged seven.

Barber, Edwin   Bridgnorth, 1841-1856

Watchmaker. On the 1841 Census he is returned as aged 28, Watchmaker of North Gate. The Return of 1851 gives a fuller picture: 'Edwin Barber, marr: aged 35, Watchmaker, Whitburn Street. born Farnworth, Staffs.
Mary Barber, wife, aged 37, born Farnworth, Staffs.
Sarah Barber, dau: age 14, born Farnworth, Staffs.
Richard Barber, son, aged 12, born Farnworth, Staffs.
Edwin Barber, Watchmaker, High Street. 1856.' (D.)

Barber, Mary   Bridgnorth, 1870-1875

Mrs. Mary Barber, watchmaker, 5 West Castle Street, 1870-1875 (D). 'Mary Barber, widow, aged 41, Jeweller, West Castle Street, born Tamworth, Warw:'. 'John Barber, brother in law, Unmarr: age 25, Watchmaker, born Tamworth, Warw:' (C, 1861). On the Census of 1871 she is listed as: 'Mary Barber, widow, aged 60, Watchmakers widow, born Tamworth, Warw:' It is of interest to note that her birthplace on the Return of 1851 is different to that which she puts on the 1861 and 1871 Returns.

Barber, John   Bridgnorth, 1861-1868

John Barber, Clock and Watchmaker, West Castle Street 1863-1869 (D). 'John Barber, Unmarried, aged 25, Watchmaker of West Castle St, born Tamworth, Warw:' (C, 1861). Brother of Edwin Barber 1841-1856.

Barnett, Jonathon   Oswestry, 1719

'1719. Pd Jonathon Barnett for looking after ye Clocke 00: 10: 00.' (Oswestry Corp. Accounts). N.B.—This would refer to the Bailey clock.

Barrett, John (Senior)   Newport, 1696-1729

Watchmaker. Children by Mary, his wife: Elizabeth 1696, and John 1698/9. His wife, Mary, died 21 March 1703/4. John Barrett died 26 October 1729. (R.)

Barrett, John Newport, 1698-1738
Clock and watchmaker. Son of John and Mary Barrett above. Baptised at Newport 8 January 1698/9. Long-case clock, narrow oak case, brass face inscribed 'Jo^h. Barrett, Newport'. 30 hour, round glass window in centre of door. Children by Millicent, his wife: Samuel 1722, bur. 1732; Charles 1728; Millicent 1724. John Barrett was buried at Newport, 26 February 1738/9. His wife died in 1756. (R.)

Barratt, Charles Shrewsbury, 1728-1807
Clock and watchmaker. Son of John and Millicent Barrett above. Bapt. Newport 12 September 1728. (R.) Apprenticed to James Evans of Shrewsbury: 'Charles Barratt hath put him Selfe an apprentice to Jas. Evans Watch & Clockmaker he giving at ye date of his Indenture. Recev^d. for the Use of the Company 5s.'. (CBR, *c.* 1747.) Admitted a Burgess of Shrewsbury 1774. In 1778 he is included as one of the Trustees of lands belonging to the Unitarian chapel in the High Street. Listed in the Poll Books as Watchmaker of Shoplatch 1774, of Swan Hill 1796, and of Dogpole 1806-1807. His death is recorded in the *Salopian Journal*: '30 Dec: 1807. A few days ago Mr. Barratt, watchmaker in this town. Died.'

Basford, Daniel Newport, 1782-1836
Clock and watchmaker, High Street 1822/3-1836 (D). Children by Jane, his wife: Elizabeth 1782, bur. 1812; Daniel 1784. A mass baptism of five of his children took place on 3 May 1796: Mary, Richard, Thomas, William, and Margaret. Mary bur. 1800. Jane baptised 1798.

Baxter, John Shrewsbury, 1812-1835
Watchmaker of Gullet-Passage. Son of Thomas Meeson Baxter, currier and Mary his wife. Admitted a Burgess of Shrewsbury 1812. Listed as a watchmaker of Gullet-Shut, in the Poll Books of 1812-1830. He married a second time at the Abbey Church in 1825, as 'John Baxter, watchmaker, of St. Chads, widower, to Mary Davies.' (R.) He died in 1835; his epitaph reads: 'On Saturday last, Mr. John Baxter, of Gullett-Passage, in this town, watchmaker, aged 54, leaving a wife and three small children wholly unprovided for.' (*SJ*, 11 March 1835.)

Biographical List of Clock and Watchmakers

Baxter, Mary   Shrewsbury, 1835-1851
'WATCH-MAKER. MARY BAXTER. (widow of the late John Baxter). Returns her grateful thanks to the public generally for the distinguished Favours manifested during the long period her late Husband carried on the above Business, and begs to solicit a continuance of the same. Having engaged an experienced Workman, she will be enabled to execute every Branch of the Business in a Manner worthy of that Support respectfully solicited.' (*SJ*, 1835.) 'Mary Baxter, widow of John Baxter, Gullet-Passage. buried 29. June. 1851, aged 61.' (R.)

Baxter, William   Shrewsbury, 1787–1828
Watchmaker of Gullet-Shut, son of Elisha Baxter, baptised 1787 (R). Possibly a relation of John Baxter above, or even working with him, their address being the same in 1828. (D).

Benbow, John   Northwood (nr. Prees), 1696–1806
Clock and watchmaker. Baillie puts his birth as 1696, but a search of the local church failed to bring it to light. However, the search did prove that he married in 1727 Jane Millington at Prees, and that their son, Thomas was baptised there in 1739 (R). He attended the church clock at Prees for a yearly salary of 10 shillings from at least 1772 until 1781. Old John Benbow's burial is entered in the registers of Prees, and his age is given as 106; the entry refers to him as 'The Well-known watchmaker'. His epitaph appears in the *Salopian Journal* of 12 March 1806:

'A few days ago, at Northwood, in the parish of Prees, in this county, Mr. John Benbow, clock and watchmaker, at the advanced age of 107. He was of the same family as the famous Admiral Benbow; was universally esteemed for his integrity and ingenuity; and what is very surprising, he executed the most intricate branches of his profession till within a few years of his death, and retained his mental faculties unimpaired to his latest moments. He lived in three centuries, and a son, and several great-grandchildren resided with him at the time of his decease. He was remarkable for sobriety, early rising, and retiring soon to rest; the liquor to which he was most partial was treacle beer. About three years ago, his taylor brought him a new coat, which he examined, and perceiving a velvet collar had been forgotten, was so irritated, that he walked to Whitchurch, the distance of seven miles, to buy one, and returned home in a few hours, to the great astonishment of his family.'

Benbow, Thomas   Northwood (nr. Prees), 1739-1809
Watchmaker. Son of John Benbow (1696-1806), and his wife, Jane. Baptised at Prees in 1739. Married Jane Davies of Hodnet, at Prees in 1767. Children: William 1768, bur. 1776; Mary, 1770, bur. 1776; Thomas 1773; Jane and Hannah (twins) 1775. (R.) A watch with twin wheel verge escapement 1785, in the Ibert Collection (Baillie). Took over and attended the church clock at Prees from his father:
> '1779. 23. March. pd Thomas Benbow a years pay for Looking after the Church Clock   0-10-0.
> 1782. March. 1. pd Thomas Benbow a years sallary for Looking after the Church Clock   0-10-0.
> 1785. Pd. Tho. Benbow a Years Sallery for Looking after the Clock. 0-10-0.'

Mr. Benbow (no initial) looks after the clock until 1790. He survived his father by only three years, being buried at Prees, aged 69, on 10 September 1809 (R).

Benbow, Thomas   Newport, 1775-1797
Thomas Benbow, Newport (Salop.). 1778-1800 (Baillie). Noted 1775 (SRO, Newport Corp. Deeds). Thomas Benbow, watchmaker 1789. Newport. (*UBD*.) Late of Newport, now of Carnaby St, St. James, Westminster, watchmaker. March 1797. (SRO, Newport Corp. Deeds.)

Benbow, Thomas   Wellington, 1773-1833
Son of Thomas Benbow (1739-1809) and Jane his wife. Baptised at Prees 21 May 1773 (R). Watchmaker of New Street. 1828-1833 (D).
> 'On the 26th ult. aged 59. Mr. Thomas Benbow, watchmaker, of Wellington, in this county.' (*SJ*, 10 July 1833.)

Bickerton, George   Ellesmere, 1789-1822
Clock and Watchmaker (*UBD* 1789). Clock & Watchmaker, High Street, 1822/3 (D).

Biographical List of Clock and Watchmakers 33

Bickerton, Thomas Osten Bridgnorth, 1870-1900
Clock and Watchmaker, 44 High Street, 1870-1875 (D).
38, High Street. 1879-1900. (D)

Birchal, George Shrewsbury, 1722-1738
Watchmaker. Married Margaret Gorsuch in 1722, at St. Mary's, Shrewsbury, who died in childbirth 1724, when their daughter, Margaret was born. He married secondly, in 1725, Elizabeth Lateward, by whom he had two daughters: Hannah 1726, and Elizabeth in 1728 (R). He took as an apprentice in 1729, Thomas Traunter of Welshpool (q.v.). George Birchal was buried at St. Chad's in 1738. His will proved in the Peculiar of St. Mary's appears in another part of this book.

Birchall, Samuel Oswestry, 1800
Clockmaker. 'John, son of Samuel Birchall, clockmaker & Sarah, born 28.4. bapt: 18.5. 1800.' (R.)

Blakeway, Thomas Rushbury, 1724-1805
Clockmaker. Son of Thomas and Hannah Blakeway of Rushbury. Bapt. 1724. Brother to Chas. Blakeway, clockmaker of Albrighton. Thomas Blakeway was a maker of turret-clocks. An example, still working, is to be seen in the church of St. Michael, Wentnor, made in 1784; another also still keeping good time, is in the tower of St. Edith's, at Church Pulverbatch. This clock is dated 1775. A brass plaque, shaped like a miniature Long-case clock dial, hangs on the wall at the back of the church, and is inscribed: 'In the year 1775 This Clock was gratuitously given to the Parish of Church Pulverbatch by Philip Jandrell Gent. of the said Paris. Makers Name Tho^S Blakeway. Rushbury No. 47'.

In 1794 he made a clock for the parish church of Churchstoke, of which perhaps the only remnant is an old clockface of lead on cast iron which was found during restoration work in 1950, and now hangs in the belfry.

Several mentions of this clockmaker are to be found in the Churchwarden's accounts of All Saints, Berrington:

'1765. Pd John Maul for taking the clock to be mended   0.2.6
   A journey to the Clockmaker's   0.2.6
   Pd Jona^S Rudge for going for the clock   0.6.0

Pd the clockmaker 3.3.0
1773. Pd Mr. Blakeway for ye Clock 0.11.0
1792. Mr. Blakeway's Bill for cleaning the Clock and his Journey 0.7.6
1796. a New Church Clock 24.0.0
1797. Paid Blakeway for mending ye Clock 0.5.0
1801. Mr. Blakeway repairing the Clock 0.2.10.'
Children by Etheldra his wife: John 1761, bur. 1761; Priscilla 1763, bur. 1773; Thomas 1765; John 1768; Richard 1771. Of these two sons became clockmakers: John at Rushbury, and Thomas jun. at Wenlock. In the Rushbury registers is entered the burial of Thomas Blakeway on 9 June 1805.

Blakeway, Charles   Albrighton, 1749–1809
Clockmaker. Son of Thomas and Hannah Blakeway of Rushbury. Bapt. 1749 (R). Brother to Thomas Blakeway, clockmaker of Rushbury (1724–1809). In 1770 he married Elizabeth Barney of Albrighton, at which event he was described as being of the parish of Dudley, co. Worcestershire. Children by Elizabeth his wife: Priscilla 1771, who married her cousin, John Blakeway, clockmaker of Rushbury in 1796; and Elizabeth 1773, who married John Hawkes in 1802. Baillie quotes an advertisement of Charles Blakeway: 'Wheels cut for clockmakers at 6d per set and Dyal Plates engraved at 2s. 6d.' His wife, Elizabeth, was buried at the age of 70, in 1808. Charles Blakeway died in 1809, aged fifty-nine.

Blakeway, Charles   Shifnal, 1789
Clock and watchmaker. Shifnal 1789 (*UBD*). Probably the same man as above.

Blakeway, John   Rushbury, 1768–1796
Clockmaker. Son of Thomas Blakeway, clockmaker of Rushbury and Etheldra his wife. Bapt. 1768. Brother to Thomas Blakeway, clockmaker of Wenlock. Married his cousin, Priscilla, dau. of Charles Blakeway, clockmaker of Albrighton, in 1796 (R). Long-case clock, 8 day, brass face.

Blakeway, Thomas   Much Wenlock, 1765–1795
Clock and watchmaker. Son of Thomas and Etheldra Blakeway

Biographical List of Clock and Watchmakers 35

of Rushbury. Bapt. 1765 (R). Child by Mary his wife: Frances, bapt. 1794. His wife Mary died 1795 (R). Mentioned as a clockmaker in 1789 (*UBD*).

Blakeway, Thomas   Broseley, 1836-1850
Another clockmaker of the numerous Blakeway family of Rushbury. 'Thomas Blakeway, aged 65, Watchmaker of the High Street. born in this county.' (Census Return 1841.) Listed in *Directories* 1836-1850.

Bond, Henry Charles   Bishopscastle, 1851-1856
Watchmaker, Church Street. Appears in *Directories* 1851-1856. Entered on the Census Return for Bishopscastle in 1851: 'Henry, Charles, Bond, aged 20, unmarried, Church St. Watchmaker. born Kew, Surrey.' Moved to Church Stretton.

Bond, Henry Charles   Church Stretton, 1868-1891
Watchmaker. Moved from Bishopscastle (see above). Shown as watchmaker in *Directories* 1868-1891, and during the period 1885-1891 additionally as Registrar of Marriages. 'Henry. C. Bond. Unmarr: age 38, Watch & Clockmaker, born Kew, Surrey.' (C, 1871.)

Boucker (Bowker?), John   Market Drayton, 1849-1850
Watchmaker of Cheshire Street. 1849-1850 (D).

Bowker, George   Market Drayton, 1851-1875
Watchmaker, Cheshire St. 1851-1875 (D). Appears in the Census Returns of 1851, 1861, and 1871 at Market Drayton, where he was born *c.* 1814. His wife, Ann, was born at Moreton Say *c.* 1822. Children by Ann his wife: Mary Ann *c.* 1844; Leah *c.* 1846; Alice *c.* 1848; Thomas *c.* 1850; George *c.* 1853; Elizabeth *c.* 1856; Mercy *c.* 1858; and Harry *c.* 1861. All born at Drayton.

Bowley, W. Shrewsbury, 1818
Engraver, barometer and thermometer maker. Opposite Mr. Blunt's Wyle Cop. (Advert., *SJ*, 1818.)

Bowyer, William Newport, 1799–1806
Watchmaker. Married Martha Spender 1799. Children: William bur. 1800; John bur. 1803; Thomas bapt. 1806 (R).

Bradshaw, Ellis Shrewsbury, 1670–1707
Watchmaker of St. Alkmund's parish. Married Mary Davies 1670. Son John bapt. 1671 (R). He was buried at St. Chad's: '12. May 1707. Elis Bradshaw, watchMaker, A prisoner buried.' (R). The reason for his imprisonment is not known.

Bradshaw, George Whitchurch, 1826–1856
Clock and watchmaker, engraver and jeweller. Pepper Alley 1828–1846; High Street 1849–1856 (D). He employed two apprentices in 1826: John Woolrich and John Furber. Long-case clock, 8 day, enamel dial, oak and mahogany banded case. The 1841 Census Returns put his age as 45; his wife, Elizabeth, as 35; and his children: Hana 8; Sarah 5; Emily 4; George 1.

Bradshaw, George Ellesmere, 1836
Watchmaker. High Street. 1836 (D). Could be a second shop of the above.

Bradshaw, Joseph Whitchurch, 1851–1879
Watchmaker and jeweller. 6, High Street. 1851–1879 (D). Round wall-clock, white face.

Brady, John Newport, 1841
Clockmaker. Appears on the Census Return for 1841: John Brady, age 60, Clockmaker, of Eccle's Yard.

Brisbourn, Peter Roddington, 1723–1793
Little is known of this clockmaker. A long-case clock of his manufacture has been noted, *c*. 1750, inscribed: 'Peter

Brisburne fecit'. Single hand. He was born in 1723, son of
Peter and Margaret Brisbourn. Buried in 1793, aged 70.

Brodhurst, Walter   **Ludlow, 1710**
Clock and watchmaker. Admitted a Freemaster of the Guild of Hammermen, Ludlow, 1710, for the sum of five pounds.

Brodshawe, Adam (1)   **Shrewsbury, 1582–1622**
Clockmaker and blacksmith. Child by his first wife, Elizabeth: Adam 1582. His wife, Elizabeth, died in the March of 1594/5. On 17 November 1596 his son Richard was baptised at St. Julian's, and 10 days later at St. Mary's he married Margerye Drewrie. Children by his second wife: Elizabeth 1602; Quads (Adam, Margery, and two boys) 1604, all died; Jane 1606, bur. 1607; John bur. 1611. His second wife, Margerye, died 1624 (R). He made a new clock and chimes for St. Oswald's, Oswestry in 1594. Also entered into an agreement for its upkeep for the rest of his natural life: 'Memorandum that the 21st of Januarij 1594 it was Agreed By the Consent [of the, *deleted*] of the parishe wth Adam Brodshowe Blacksmith of the towne of Salope for mackinge of A newe Clocke And Chimes All new except iij whilles Allso he is to macke j Dyall wth such nesesaries thereunto belonginge as aney Sufficient worckman shall thinke required, further the said Adam doth promis and Covenant to repair the same from tyme to tyme Duringe his naturall life, painge him his Charges with such resonable recompence as the best gent and wardens for the present tyme shall thinke fitt. In Consideration heere of the parishe is to payhim the some of Tenne pounds I say xli provided if this Adam be alouser by the mackinge of the premissis and that (it may appeare, *interlined*) Aparant unto the parishe then he is to have the some of XXs more for his labor for the suficient mackinge And sentinge up of the Clocke And Chimes with the Diall mr. Richard lloyd of Aston And his Brother, mr. Richard Kyffin Lloyd hath passed ther words for the Accomplishment of promissis whereof at this present he Rd in part xls and is to have at the sentinge up of the Clocke And deiall iiij li And the rest at the setting of the Chimes (in another hand) Mem that the said Adam upon the

viijth of Decembr 1598 haith undertaken that he shall and
will upon viij daes warninge make his present apaier to this
tower from tim to tim During his lif if sicknes Does not let him
thereof and in every of thoes times sufficiently repaier the said
chimes and clocke and Dieall the charge of the said severall
yurnes to be bornne by the parish and considerarcion for his
paines to be aloued by the said parich singd XX Adam Brodshowe.

  Gryff. Kyffin.   Th. Evance
  Tho. Staney   hughe yale.'

Several entries in the Oswestry churchwarden's accounts show
that Adam Brodshowe kept to his part of the agreement,
travelling to Oswestry to attend the clock. In 1608 for example:
  'It to Evan ap Thomas labourer for a Jorney to Salop
  Towne and for the hire of his horse to fetch Adam Brodshiaw
  hither to amend the clocke xviijd.'

Then appears in the next iten the clockmaker's charge:
  'It To Adam Brodshiawe for mending of the Clocks and
  Cheymes Xls.'

Adam Brodshawe died on 22 March 1621/2 (St. Julian's R).

Brodshawe, Adam   Shrewsbury, 1582-1639
Clockmaker. Son of Adam Brodshawe and Elizabeth his first
wife (above). Bapt. 1582. Married Jone Rowley in 1610, at
St. Julian's. Repaired the clock of St. Peter's Cound, in 1625,
for which he was paid One shilling and eightpence (CW).-Adam
Brodshawe II was buried at St. Julian's in 1639. His wife, Jone,
following him in 1651 (R).

Brown, Henry   Shrewsbury, 1841-1895
Watchmaker. (*See* Evans and Brown.) Census Return of 1851
notes his age as 33, and his place of birth Newtown, Montg.
Watchmaker, Coleham Head, age 20 (C, 1841).

Brown, John   Wem, 1828-1834
Watchmaker of the High Street. 1828-1834 (D).

Brown, Joseph   Oakengates, 1871
Joseph Brown, Watchmaker. 1871 (D). Watchmaker of Market

St, aged 24, born Coventry, Warks. Catherine Brown, wife, age 22, also born at Coventry (C, 1871).

Brown, Thomas   **Ruyton Xl Towns, 1772-81**
Clockmaker. Will dated 1772. Buried 1781.

Bryan, William   **Ludlow, 1755**
Watchmaker, noted on Ludlow conveyance of property. 6 May 1755 (SRO).

Bullock, Edmund   **Ellesmere, 1708-1734**
Clockmaker. Attended the clock of St. Mary's, Ellesmere, from 1725 until his death.

> '1725. pd Edm Bullock for Looking after ye Clock & Chimes 1. 15. 1.'

> '1726. pd Edm Bullock for Work done at ye Diall 0. 15. 11.'

(CW). The sundial to which the last entry refers is the one in St. Mary's churchyard, which is inscribed with the names of the four churchwardens followed by that of 'Edm Bullock. Ellesmere, fecit 1726.' In 1718 he did a repair job for the Middletons of Chirk Castle:

> '1718. July 11. Pd Mr. Edmond Bullocke of Elsmere, for comeing to put ye spring clock in order yt is in Madm Mary Myddleton's Chamber. 0. 2. 6.' (CCA.)

A very handsome eight-day, long-case striking clock, with an oak case decorated with Chinese figures on a ground of tortoiseshell lacquer, and inscribed 'Edmund Bullock, Ellesmere. 303.' is in the possession of the Howard family. It stands 8 feet 8 inches high (without the spires). It is reputed to have been made to the order of the East India Company, or for one of its officers. Tradition has it that it was, at one period, in India, the property of Hugh Howard, an ancestor of the present owner. It is illustrated in Cescinsky and Webster's *English Domestic Clocks,* 1913. Children by Sarah, his wife: Sarah 1708; Mary 1709; Thomas 1713; Jeremiah 1716, bur. 1716; Martha 1717; Richard and Hannah (twins) 1719; Jeremiah 1873; John bur. 1725 (R). Edmund Bullock died 16 June

1734; the administration of his will was granted to his widow, Sarah. An inventory of his goods amounted to £138 8s. 2d.

Bullock, Jeremiah Ellesmere, 1723–1780
Clockmaker. Son of above, Edmund Bullock, clockmaker, and Sarah, his wife. Bapt. 1723. Brother of Richard Bullock, clockmaker. Married Ann Jackson 1771. Child: Edward Adams Bullock 1772. Jeremiah Bullock was buried at Ellesmere 19 June 1780.

Bullock, Richard Ellesmere, 1719–1797
Clockmaker. Son of Edmund Bullock, clockmaker, and Sarah his wife. Bapt. 1719. Brother of Jeremiah Bullock, clockmaker. Long-case clock, 8 day, brass dial, numbered 220, mahogany banded oak case. This maker was reputed to make Turret clocks. Richard Bullock, watchmaker. 1789 (*UBD*). He died at Ellesmere where he was buried: '3.11.1797. Richard Bullock of Ellesmere, clockmaker. aged 78. buried' (R).

Bullock & Davies Ellesmere, 1789
Clock in Cockshutt church bears their name, pointing to a partnership. Clock made 1789, both the clock and dial bearing the date. The gift of Mr. Roger Jones of London. Partners—possibly Richard Bullock and Edward Davies.

Bullock, Edward Oswestry, 1732–1733
Clockmaker. Attended the clock of St. Chad's, Prees. 1732: 'pd Edward Bullock towards ye Chines 10. 0. 0. Given to his son and man 0. 1. 0. pd for Clening the Clock 0.2.6. pd att setting them up 0. 6. 1.' 1733: 'pd Mr. Dicken for Drawing Mr. Bullock's bond 0. 3. 6.'

Bullock, William Wootton, Oswestry, 1742
Clockmaker of Wootton, buried at Aston. 1742 (R). Long-case clock, 8 day, brass dial, one hand.

Bullock, Gilbert Bishopscastle, 1724–1773
Clockmaker. Payments were made by the Bishops Castle Cor-

poration for his attention to the Tower clock, as appears in their Chamberlain's accounts over the years 1727-1737. Married Catherine Davies in 1724. Children: Elizabeth 1726; Richard 1728, bur. 1731; Anne 1732. Gilbert Bullock was buried 1 May 1773.

Burnett, Charles   Ludlow, 1729-1743
Clockmaker. Long-case clock, 8 day, brass face, oak case, brass finials. Married Sarah Lugg in 1729. Children: James 1730; Elizabeth 1732, bur. 1743; Sarah 1733.

Burroughs, James   Ironbridge, 1828-1848
                                              Dawley, 1849-1868
Clock and watchmaker. Ironbridge, 1828-1848 (D). Long-case clock, 8 day, painted dial, inscribed 'Jas. Burroughs. Ironbridge'. Also mahogany cased 'Act of Parliament' clock, inscribed similarly. Children by Ann, his wife: Sarah c. 1826; John 1829; Thomas James c. 1831; Rowland c. 1833; Elizabeth 1835; Joseph c. 1838. Moved from Ironbridge to Dawley Magna by 1849, where he is shown as from that year until 1868 in directories. The 1851 Census returns him as aged 56, a Clock & Watchmaker of Dawley Green, born at Tilstock. His wife, Ann, aged 49, born Middleton, Salop. A daughter, Sarah, aged 25, Bonnet-maker, born Ironbridge, as were all his children. John, his eldest son, aged 22, Clockmaker & Watchmaker. His second son, Thomas James, aged 20, a Cabinet-maker by trade. Rowland, aged 18, Clock & Watchmaker. Elizabeth, aged 16. Joseph, aged 13. When the 1861 Census was taken, the chicks had flown the coop, and only James, now aged 65, a Clock & Watchmaker in the High Street, and his wife, Ann, are returned.

Burroughs, Rowland Kitson   Ironbridge, 1833-1875
                                              Dawley, 1870-1875
Clock and watchmaker. Son of James and Ann Burroughs. Bapt. at Ironbridge c. 1833. Still living with his parents at Dawley in 1851 (C). Watch and clockmaker, Salop Road 1863-1875 (D). Second shop in the High Street, Dawley, 1870-1875 (D).

Butler, Henry Wem, 1840-1875
Watch and clockmaker, Mill St. 1840 (D). High Street 1849-1875 (D). In the Census of 1851 he is returned as: age 35, Clock & Watchmaker of New Street, born Whitchurch. His wife, Mary Ann Butler, aged 38, born Baschurch. A decade later in the 1861 Census, he still appears at the High Street address, but his first wife had died, and the name of his second wife is given as Rose H. Butler, aged 24, born Clive. Also two sons: Josiah, aged 4, born Hadnall; Alfred, aged 1, born Wem. In the 1871 Census three more sons and a daughter have been added to the family, all born at Wem: Walter, aged 7; Clement, aged 4; Herbert, aged 3; and Julia, aged one.

Byn, alias Byrd, alias Burd, Richard Wenlock, 1642-1659
'Paid Richard Burd of Wenlocke for keepinge St. Chaddes Chimes in order this yeare. 0. 10. 0.' (SCA, 1642.)
'Mr. Waltall, I pray you pay to Richard Byn, of Wenlock, the sum of tenne shillings for keeping the chymes in order this yeare last past. March 19th. 1645.' (SCA.)
'It. pd Richard Byrd of Wenlocke for a Clocke and dyall £7. 0. 0. It. pd him for refreshing the Son Dyall 17s. 6d.' (CW, St. Julians', Salop.)

Calcott, Richard Cotton, nr. Wem, 1719-1784
Clockmaker. Son of John and Lyddia Calcott. Bapt. at Edstaston 20 April 1719 (R). Attended the clock and chimes of St. Chad's, Prees, 1746-1763, at a salary of 15 shillings per annum. Married 1743, at Wem, Jane Wood (R). Child: Hannah, bapt. 1749. Richard Calcott was bur. at Edstaston 17 November 1784. His widow, Jane was bur. 1793.

Calcott, John Cotton, nr. Wem, 1753-1830
Clock and watchmaker. Son of Arthur and Jane Calcott of Cotton. Bapt. 23 April 1753 at Edstaston. Nephew of Richard Calcott, clockmaker (1719-1784). Married Sarah Bradley, at Wem in 1776. Child: John, bapt. at Edstaston 1777.
'Sept. 28. WATCH LOST: On Monday Night, the 19. of September, either in Tilstock, or between there and the new

windmill on Cotonwood: A SILVER HUNTING WATCH, with a small Glass and Name on the Dial instead of Figures, on the outer circle the Name of Samuel Tebb, and between the Circles (Furnilees). Makers Name, John Callcott, Coton. No: 2842; Whoever has found the said watch, and will bring it to the said JOHN CALLCOTT, shall receive One Guinea Reward.' (Advert., *SJ*, 1813.) His watches appear to have a habit of being lost, one was advertised for in 1810: 'A SILVER WATCH, marked John Calcott, Cotton, No: 2594.' And once again: 'LOST. On Thursday, the 26th Inst. between Whixall Moss & Whixall Chapel. A Silver Watch, MakerS Name John Calcott. Cotton. No: 2525. Whoever has found the said watch, and will bring it to Mr. John Calcott, shall receive a Reward of Half-A-Guinea; whosoever shall conceal the said watch after the Date of this Notice, will be punished to the utmost Rigour of the Law. February. 1st. 1817' (*SJ*).

'On the 5th inst. at Cotton, near Wem, after a long & protracted illness, which he bore with true christian fortitude & pious resignation, Mr. John Callcott, clock & Watchmaker, in his 78th year: as a neighbour his acquaintance was generally courted & esteemed, & as a mechanic his genius was of more than an ordinary class' (*SJ*, 17 November 1830).

Calcott, John   Cotton, nr. Wem, 1777-1853
Clock and Watchmaker. Son of John Callcott, clockmaker (1753-1830) and Sarah, his wife. Bapt. 21 Sept. 1777, at Edstaston (R). Long-case clock, painted dial, inscribed: 'John Calcott. Cotton.' Married Anne Batho at Wem, 18 November 1801. Child: Mary, bapt. at Edaston 1802. 'John Calcott, marr: age 73, Clock & Watchmaker. born Cotton. Ann Callcott, wife, aged 68, born Tilstock. Martha Chesters, grand-dau: age 10, Samuel Chesters, grandson, aged 8.' (C, 1851.) John Callcott was buried at Edstaston on 16 February 1853, aged 75.

Callcott, John   Prees, 1824-1833
Clock and watchmaker. Made a sundial for St. Chad's, Prees, in 1824. Married Anne Hadley at Prees in 1825. Children: Mary 1825; Sarah 1829; Ann 1833 (R).

Calcott, John   Wem, 1832-1850
Watchmaker. High Street, Wem. 1832-1850 (D).

Calcott, John   Whitchurch, 1840-1851
Watch and clockmaker, High Street. 1840-1851 (D). Shown as Thomas in 1851 *Directory*.

Camell, William   Ludlow, 1778
'5th May 1778. Then William Camell was admitted a Free-master of this Company to the Trade of a Clock and Watchmaker he foregoing down a ffine of three pounds (two pounds being abated him out of his ffine of five pounds) he being a Forreigner which he accordingly did.   William Camell.' (GHR.)

Campbell, Francis   Oswestry, 1822-1841
Clock and watchmaker, Cross Street 1822/3-1834 (D). An Alderman of Oswestry, and Mayor of that Town 1836. He died, aged 73, on 26 September 1841. His wife, Mary, died, aged 78, in 1845.

Campbell, Henry   Oswestry, 1871
Watchmaker, boarding with George Hayes and Sarah his wife, in Leg Street, married, aged 55. Born Hexham, Northumberland. (C, 1871.)

Campbell, Robert   Shrewsbury, n.d.
Long-case clock, painted dial, inlaid oak case, 8 day. 19th century.

Capper, John   Shrewsbury, 1578-1581
'1578. Pd to Capper for mending the clocke, which my L. his men brake, 2s.' N.B.—This entry refers to the clock of the Abbey Church, but gives no reason why the Lord's men should have broken it. (O & B, Vol. II, p. 152.)
John Capper, clerk of the Abbey Church, was hung for treason on 24 March 1581.

Careswell, Francis Shrewsbury, 1776-1823
Clock and watchmaker, Mardol Head. 1786-1822 (D). Admitted a Burgess of Shrewsbury 1795. Long-case clock, 30-hour, painted dial, inscribed 'Careswell. Salop.' Children by Priscilla, his wife: George Daniel 1776; Harriet 1782; Hannah, bur. 1785; Frances 1788, bur. 1794 (R). Advert.: 'Francis Carswell. Clock and Watchmaker, Mardol-Head. Takes this opportunity to return his most grateful Acknowledgements for the liberal Favours conferred on him since his commencement in Business; and begs leave to inform his Friends and the Public, that, in Addition to his Clock and Watch Business, he has laid in a General Assortment of Articles in the Jewellery Line, as well as an entire new Assortment of Ladies Head Ornaments, with all kinds of Japanned Goods, &c &c. which he is determined to sell on the very lowest Terms. Genteel Lodgings. Shrewsbury, 9th September, 1806.' (*SJ*, Advert. 1806.)
'On Thursday night last, aged 72, Mr. Francis Careswell, watchmaker, of Mardol Head, in this town.' (*SJ*, 1823.) His wife Priscilla died two months before him.

Carswell & Harper Shrewsbury, n.d.
Of Shrewsbury. Late 18th century (Baillie).

Cartwright, Thomas Wem, 1861-1871
Watchmaker, Noble Street, aged 18 (lives with widowed mother), born Wem. (C, 1861.) Shown in the 1871 Census Return as unmarried, aged 28, now of Mill Street.

Cartwright, William Ellesmere, 1758-1770
Clockmaker. '1758. Paid Jno Joyce & Wm Cartwright one year's Salary for Looking after the Clock due at Lady Day last. 3. 0. 0.'
'1768. pd Wm Cartwright for Work done at ye Chimes. 0. 5. 0.'
'March 27th. 1770. They agreed at a Vestry meeting this day that William Cartwright be allowed four pounds pr. anm to look after the Church Clock, in consideration of the said William Cartwright finding ropes, pulles, oil, wyre and all other Materials

to keep the Clock & Chimes in proper order, without adding any additional Expense to the parish.' (CW Accounts, Ellesmere.)

Cetti, Paul & Co. Wellington, 1841–1879
Watchmaker, jeweller, furniture dealer, 25, New Street. Paul Cetti, aged 20, jeweller, New Street (C, 1841). Paul Cetti & Co., New Street. 1868–1879 (D). In partnership with Goetuno Del Vecchio, under the name of Del Vecchio & Cetti, jewellers, cutlers, furniture brokers, in 1856 (D). In 1841 the Census shows on the same premises: Dominic Dotti, Jeweller; Goetuno Del Vecchio; Samuel Frankel; and Paul Cetti. Various combinations of partnerships between these four persons appear at Wellington in the second half of the 19th century. Children by Frances, his wife: John Cetti, born c. 1852 at Wellington; Margaret Cetti, born c. 1853 at Wellington; Paul Cetti, born c. 1856; Fanny Cetti, born c. 1858. In 1861 his nephew, John Cetti, aged 22, and a watchmaker, lived with him. He and his wife and nephew were all born in Italy.

Charlton, Edmund Lechmere Ludlow, 1830
Watchmaker. Admitted 5 July 1830 to the Freedom of the Company of Hammermen, Ludlow (CHR).

Chune, Thomas Shifnal, 1781–1796
Clockmaker. Children by Anne, his wife: Eleanor, 1781, bur. 1782; Ann, bur. 1781; Petrach 1783, bur. 1785; Ann Harriot 1786; Mary 1796 (R).

Churton, Joseph Whitchurch, 1799–1836
Clock and watchmaker, High Street. 1828–1836 (D). Children by Ann, his wife: Mary 1799; Joseph 1802; Elizabeth 1805; John 1807 (R).

Churton, Joseph Whitchurch, 1802–1841
Prees, 1851–1863
Clock and watchmaker. Son of Joseph and Ann Churton above. Bapt. 6 June 1802 at Whitchurch. Shown on the 1841 Census at 35 High Street, Whitchurch, as a Clock & Watchmaker, with Mary, his wife, aged 40; dau., Frances, aged 12; son, Joseph,

Biographical List of Clock and Watchmakers

aged 9; dau., Ann, aged five. By 1851 he had moved to Prees, and is listed in the directories from 1851 until 1863 as both a clockmaker and landlord of the *New* Inn.

Clarke, Elijah   Oswestry, 1871
Watchmaker of Gatacre Place, Oswestry, aged 28, born Dorsetshire. Wife: Mary, aged 25, born Kent; two children: Evelyn, a daughter, aged 2, and Hedley a son, aged 9 months, both born in Kent.

Clarke, Jane   Wellington, *c.* 1825
Watchmaker. 'Jane Clarke, Wellington. *ca.* 1825. Watch in Shrewsbury Museum.' (Baillie.)

Clench, Richard   Ludlow, 1582–1627
Attended the clock and chimes of St. Lawrence's, Ludlow 1625–1626. 'payd Richard Clench the elder ffor keping the Chimes and Cloke ffor the yeare past 1625 by order of the sidmen upon his peticcon Vs.' also 'ffor his wages ffor the whole year vjs. viijd.' (CW.) Children by Johan, his wife: nine in number over the years 1583–1606. He was buried at Ludlow 20 November 1627.

Clevely, Thomas   Shrewsbury, 1770–1778
Thomas Clevely, Shrewsbury. an. 1778. Watch. (Baillie.) Watch movement: 'Clevely Salop. 1067' on back, metal face. *c.* 1770 (CMH).

Clifton, Cuthbert   Shrewsbury, 1635–1642
Watchmaker. Married Elizabeth Hurst, widow in 1635. Children: Benjamin 1638; Jane 1640; William, bur. 1642. Of St. Alkmund's parish (R).

Collier, –   Newport, 1787
– Collier, Newport (Salop.) 1787. Clock and watchmaker (Baillie.)

Cooper, Benjamin   Whitchurch, n.d.
Benjamin Cooper, watch and clockmaker, Brownlow Street. (Britten.)

Cooper, Charles   Whitchurch, 1770-1774
Clockmaker. Attended the clock of St. Alkmund's, Whitchurch. 1770. 'Mr. Chas. Cooper's bill 4. 0. 0.' 1771. 'To Mr. Charles Cooper's bill 1. 15. 9.' 1774. 'Oct. 17. Mr. Cooper for a new wheel & repairing the Clock 0. 15. 0.' 1774. 'D^O Half a years Salary looking after the Clock 0. 15. 0. (CW).

Cooper, Mr.   Whitchurch, 1742-1743
Clockmaker. Attended the clock of St. Chad's Prees. 1742. 'Pd. Mr. Cooper for the Chimes 6. 0. 0.' 1743. 'Paid a Mefsenger to take the Pendelant of Clock to Whitchurch 0. 0. 3.' (CW, Prees.)

Cooper, Joseph   Whitchurch, 1765-1773
Clockmaker. Attended the clock of St. Chad's, Prees. 1742. Whitchurch (Salop.) 1773. (Baillie.) Joseph Cooper, Whitchurch (Salop.) 1765. (Britten.)

Cooper, Joseph   Shrewsbury, 1714
Joseph Cooper, 1714. Clockmaker. (Baillie.)

Cooper, Thomas   Newport, 1783-1789
Clockmaker. 1789 (*UBD*). Married Ann Allen 1783. Children: Ann 1784; Anthony 1788. Thomas Cooper was buried 14 October 1789.

Corken, Archibald   Oswestry, 1849-1852
Clock and watchmaker of Cross Street. 1849-1850 (D). At 7 Church Street 1852 (D). The 1851 Census Return shows him as Archibald Corken, married, aged 36, Clock and Watchmaker employing 3 men. Cross Street. born Scotland. Mary Ann Corken, wife, aged 34, born Chester. Archibald Corken, son, aged 9, born at Oswestry. Ellin Donat Corken, dau., age 4, born Oswestry. Mary Ann Corken, dau., aged 1, born Oswestry (C).

Corvehill, William   Much Wenlock 1546
A very learned monk of St. Milburge's Priory, Much Wenlock, who amongst many other crafts was skilled in the making of clocks. There is mention of a 'horologium' at the priory as early

as 1233 when Henry II gave four oaks towards the building of the tower to hold it. (Close Rolls, 1231-4, p. 225.)
'1546. 26. May. Bur(ie)d out of Tow tenemts in Mardfold Street next St. Owen's well. Sr. Wm. Corvehill Priest of the Service of o^r lady in this Ch(urch) which 2 tent^s belong^d to the service he had them in his Occupac'on in pt. of his wages wch was viij Mks and the sd Houses in an ov'plus. He was well skilled in Geometry not by Speculation but by Experience could make Organs, Clock and Chimes. In Kerving in Masonry and Silk Weaving and painting, & coud make all Instrum^{t}s of Musick & was a very patient & Gud Man borne in this Borowe sometyme Monk in the Monestery. All this country had a great loss of Sr. Wm. for he was a good Bellfounder & Maker of the frames.' (Bodl. Lib., Gough, Salop., 15.)

Cotterill, Henry Shrewsbury, 1875
Watchmaker. 'Henry Cotterill, 22, Princess Street, Shrewsbury. Manufacturing and Working Jeweller, Electro-plater & Gilder. 12 years principal practical man to Messrs. Sloane & Carter of Birmingham. Has now a varied assortment of all kinds of Jewellery, watches, clocks, spectacles, and eye glasses, of every description. N.B.—Repairs by post or rail promptly attended to, and returned carriage paid.' (Advert., *SJ*, 29 December 1875.) Watchpaper: 'H. Cotterill, Practical Watchmaker, Goldsmith, & Jeweller. 3, Mardol Head. Shrewsbury. Wedding Rings. Every description of watch carefully repaired & its Performance guaranteed' (CHM). This firm was later carried on by his wife, and then his son, Henry.

Crisp, Edward Wellington, 1840-1841
E. M. Crisp, watch and clockmaker, Market Place. 1840 (D). 'Edward Crisp, aged 22, Watchmaker, of New Street' (C, 1841).

Cross, William Shrewsbury, 1837-1843
'14.8.1837. William Cross, clockmaker of Frankwell & Mary Ridgway of the same, marr:' (St. Geo. R). 'William Crosse, age 30, Clockmaker, Chester St. son—William, age 3, Matthew age 1' (C, 1841). '25.2.1843. Walter Cross, son of William Cross, Clockmaker & Mary his wife, bur: and infant. ABODE—

GAOL' (St. Mich. R). On 28 February 1843, Mary Cross was committed to gaol, charged with stealing at Betton, an umbrella, valued one shilling, for which she was sentenced to three months' imprisonment, to HARD LABOUR. While awaiting trial her son was born, and unfortunately died.

Cross, William   **Ellesmere, 1841-1863**
Clock and watchmaker of Cross St. 1849-1863 (D). 'William Cross of Cross St, aged 60, marr: Clockmaker. born Wrexham. Sarah Cross, wife, age 67, born Baschurch, Salop. Sarah Cross, dau: age 22, Unmarr: Dressmaker, born Ellesmere' (C, 1851). By the next Census, 1861, his dau. had marr.: Sarah Thomas, dau: age 32, Dressmaker, John Thomas, son-in-law, age 27, Sawyer. Elizabeth Thomas, grand-dau:, age 6 months (C, 1861).

Crumpe, Richard   **Ludlow, 1603-1617**
'1603-1604. It'm to Crompe for keepinge the clocke & chimes for the year. vjs. viijd.'
'1607-1608. Item to William Crumpe for his wagis tendinge the Clocke & Chimes and Kepinge Clene the leds. xiijs. iiijd.'
'1608-9. payd William Crump his years wages ffor tendinge the clock & chimes & keepinge Clene the leads. xiijs. 4d.'
'to Richard Crumpe for keeping the clocke & chimes xs.' 1617 (CW, Ludlow).

Daniell, Hugh   **Ludlow, 1603-1646**
1640-1641. 'payd Hughe Daniell his yeares wages for wynding & keepinge the Clock and Chymes. 1. 0. 0.' (CW, Ludlow). Of his 10 children born between 1615 and 1629, to his wife, Anne, five died in childhood. Hugh Daniell was buried 23 March 1645/6.

Darken, Mr.   **Condover, 1682**
'payd to Mr. Darken for the Clock and dyall board the sum of 8. 15.0.' (CW, Condover).

Davies, Daniel   **Shrewsbury, 1804-1868**
Clock and watchmaker. Son of William Davies, clockmaker, and Sarah, his wife, of Shifnal. Bapt. 18 November 1804, at

Shifnal. 'On the 31st ult at the Abbey, Mr. Daniel Davies, watchmaker, Shiffnal, to Miss Mary Edwards, of the Crow Inn, Abbey Foregate, in this town' (*SJ*, 8 January 1834). Daniel Davi(e)s, watchmaker, Castle Street, Shrewsbury, 1842-46 (D). Daniel Davies, watchmaker, Mardol, 1849-1868 (D). On the Census Report of 1841 he is listed as 'Daniel Davis, age 35, Clockmaker of Raven St, Shrewsbury. Wife—Mary, aged 30. Children—Elizabeth 4; William 2; Edward 1.' In the Return of 1851 his household consists of Daniel Davies, 46, Clockmaker. Mardol, born Shifnal. Wife—Mary, 42, born Birmingham; son—William, 12 born Shrewsbury; dau:—Elizabeth, 14, born Shifnal; sons—John 7; James 3, & dau: Mary 1, all born Shrewsbury. Sister—Elizabeth 55, born Albrighton (C).

Davis, E. Shrewsbury, n.d.
Barometer maker. Mahogany cased stick barometer (Auction Sale, Wellington, 1977).

Davi(e)s, William Shifnal, 1798-1842
Clockmaker, High Street,. Shifnal 1822-1842 (D). Children by Sarah, his wife; William Henry 1798; John 1799; Anne 1802; Daniel 1804; Elizabeth *c.* 1811; Thomas *c.* 1816. All of his four sons became clock and watchmakers. 'William Davis, Watchmaker, of Horsefair, age 75, born in this county. Thomas Davies, aged 25. Elizabeth Davis age 30' (C, 1841). His obituary in the local press reads: 'On the 5th inst. at Shiffnal, in his 78th year, Mr. W. Davis, Clockmaker; a man universally respected & deeply lamented' (*SN* & *CR*, 19 March 1842).

Davis, John Shifnal, 1799-1875
Clock and watchmaker. Son of William Davis, clockmaker, and Sarah his wife. Born 9 November 1799, at Shifnal. Brother to William Henry, Daniel, and Thomas Davis, all of the same trade. Children by Sophia, his wife: Sophia *c.* 1830; Elizabeth, *c.* 1833; Georgina *c.* 1836; John *c.* 1840; Mark *c.* 1843; William *c.* 1848. Of these only John became a clockmaker. 'John Davis, Back Street, age 61, marr; Clockmaker, born Shifnal. Sophia Davis, wife, age 50, Dressmaker, born Albrighton. Sophia Davis, Dau; Unmarr: age 30, Milliner. born Wolverhamp-

ton. John Davis, son, age 21, Unmarr; Clockmaker. born Shifnal. Mark Davis, son, age 18, Pupil Teacher. born Shifnal. William Davis, son, age 21, Unmarr; Clockmaker. born Shifnal' (C, 1861). John Davis, clock and watchmaker. Back Street. 1851 (D). High Street. 1856-1875 (D).

Davis, John Shifnal, *c.* 1840-1875
Clockmaker, son of John Davis, clockmaker and Sophia his wife, born *c.* 1840, at Shifnal.

Davis, William Henry Shifnal, *c.* 1840-1875
Clock and watchmaker. Son of William Davis, clockmaker, and Sarah, his wife. Born 23 March 1798, at Shifnal. Brother to Daniel, John and Thomas Davis, all of the same trade. Married at Shifnal, Elizabeth Jenkins, 30 March 1826. Watchmaker, Dun Cow Lane. 1836 (D). 'William Davies, age 40, Park Street, clockmaker. Elizabeth Davies, age 35. Frances Davies, age 11. Nathaniel Davies, age 9. Sarah Davies, age 5' (C, 1841).

Davies, Thomas Shifnal, *c.* 1816-1843
Clockmaker. Son of William Davis, clockmaker, and Sarah his wife. Born *c.* 1816. 'Oct: 18. 1843. To Churchwardens & Others On Sale. A powerful Eight-Day Church Clock, strikes Hours and Quarters, and made upon improved principles— Apply to Thomas Davies, Church & Turret Clock Maker, Shiffnal; or Daniel Davies, Clockmaker, Castle Street, Shrewsbury' (Advert., *SJ*, 14 October 1843).

Davies, David Shrewsbury, 1795
David Davies. Shrewsbury. 1795 (Baillie).

Davies, Edward Ellesmere, 1781-1798
Clockmaker. Children by Mary, his wife: Elizabeth, bur. 1781; Thomas, bapt. 1790 (R). Edward Davies. Clockmaker. 1789 (*UBD*). Edward Davies. Ellesmere. Mid-18th century L/c clock (Baillie). '10. October. 1798. Edward Davies of Ellesmere, clockmaker. age 44. bur:' (R).

Davies, Edward Ellesmere, 1836-1841
Edward Davies, clockmaker, Church St. 1836 (D). 'Edward Davies. clockmaker, age 55' (C, 1841).

1. William Peplow (1794-1895), clockmaker of Wellington and Shifnal. Taken on his 100th birthday.

2. (*above*) John Trevor Griffith Joyce (1903-1971), watch and clockmaker of Denbigh.

3. (*left*) Norman Joyce (1891-1966), clockmaker of Whitchurch. Shown overhauling Marbury church clock in 1955, 111 years after it was made by his firm.

4. Advertisement in *Eddowes Journal*, 15 February 1860

STANDARD GOLD CHAINS.

ROBINSON'S London-made WATCHES, MARKET SQUARE, SHREWSBURY, are kept ready for use, silfully timed, and the performance guaranteed. Purchasers may have them cleaned and adjusted (if required) at the end of the first year without charge, if no other workman has had them.—Dials for Offices and Shops. House Eight-day Clocks, Time Pieces, Carriage Clocks, Alarums, &c.

MOURNING JEWELLERY.

ROBINSON, MAKER, MARKET SQUARE, SHREWSBURY, AND BISHOPSGATE WITHIN, LONDON.—BAROMETERS & THERMOMETERS, &c.

EXTENSION OF PREMISES.
15, HIGH STREET, SHREWSBURY.

HENRY ROBINSON,
GOLDSMITH AND JEWELLER,
WATCH AND CLOCK MAKER,
Nephew of the LATE Mr. E. H. Robinson, of the Market Square, Shrewsbury.

ONE of the Largest Stocks in the Provinces to choose from, and any Article exchanged if not approved. A quantity of good Second-hand Watches, by eminent Makers.

ELECTRIC CLOCKS.
REPAIRS.

Every construction of CLOCK or WATCH thoroughly understood, corrected, and repaired in a proper way, upon the premises, by separate workmen for each branch, and all are (if sound) **warranted for 12 months after repairing.**

MASONIC JEWELLERY.
CLOCKS ATTENDED by the Year.

OPTICIAN (by appointment) to the SALOP EYE AND EAR HOSPITAL.
SPECTACLES, from 1s. per pair.
PLATE AND CUTLERY ON HIRE.
15, HIGH STREET.

5. Advertisement in *Eddowes Journal*, 17 May 1871.

6. Advertisement in *Eddowes Journal*, 15 February 1860.

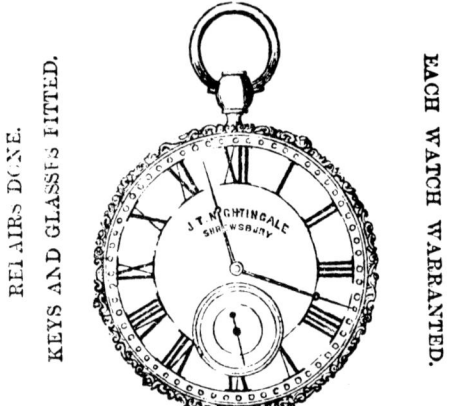

LADIES' GUARD AND GENTLEMEN'S ALBERT STANDARD GOLD CHAINS, GUARANTEED THE QUALITY REPRESENTED.

KEYS AND GLASSES FITTED. REPAIRS DONE.

EACH WATCH WARRANTED.

NIGHTINGALE'S London-made GOLD and SILVER WATCHES are offered to the public with perfect confidence as to their accurate time-keeping, and are regulated and cleaned the first two years free of charge. Each Watch marked in plain figures, subject to 5 per cent. for cash payment. Good English-made Watches at Two Pounds Ten Shillings each.

38, HIGH STREET, SHREWSBURY.
January 25th, 1860.

JOHN EVANS,
WATCH AND CLOCK MAKER,
MARKET PLACE,

BEGS to acquaint the Nobility, Gentry, and the Public in general, he is appointed (SOLELY) to vend FRIBOURG and TREYER'S genuine SNUFF and TOBACCO in SHREWSBURY. He has just received a Variety of Sorts and Mixtures, which may be had in Cannisters of every Size. J. E. embraces this Opportunity of returning his most grateful Thanks for the liberal Encouragement he has received since he commenced Business, and of assuring his Friends that any Orders they may please to confer on him, shall be attended to with the greatest Punctuality.
Sept. 28th, 1813.

7. Advertisement in *Salopian Journal*, 29 September 1813.

FRANCIS CARSWELL,
CLOCK AND WATCH MAKER,
MARDOL-HEAD.

TAKES this opportunity to return his most grateful Acknowledgements for the liberal Favours conferred on him since his Commencement in Business; and begs leave to inform his Friends and the Public, that, in Addition to his Clock and Watch Business, he has laid in a General Assortment of Articles in the JEWELLERY LINE, as well as an entire new Assortment of LADIES HEAD ORNAMENTS with all kinds of JAPANNED GOODS, &c. &c. which he is determined to sell on the very lowest Terms.
☞ Genteel Lodgings.
Shrewsbury, 9th September, 1806.

8. Advertisement in *Salopian Journal*, 10 September 1806.

William Henry Norris, Oswestry.

Robert Grosvenor, Ellesmere.

James Hanny, Shrewsbury.

Richard Giles, Junior, Shrewsbury.

William Owen, Oswestry.

Joseph Jones, Oswestry.

9. Watchpapers. (*Courtesy of Clive House Museum*)

10. Silver pair case for Verge Watch made by Thomas Gorsuch, Shrewsbury, c.1710. (*Courtesy of Merseyside County Museums*)

11. Silver pair case for Verge Watch made by Thomas Gorsuch, Shrewsbury, c.1710. (*Courtesy of Merseyside County Museums*)

12. (*opposite*) Silver pair case for Verge Watch made by Thomas Gorsuch, Shrewsbur c.1710. (*Courtesy of Merseyside County Museums*)

13. Silver pair case for Verge Watch made by Thomas Gorsuch, Shrewsbury, c.1710.
(*Courtesy of Merseyside County Museums*)

14. Long-case Clock made by Francis Arkenstall of Market Drayton, owned by Roy Betterly, Washington, U.S.A.

AN ENGLISH LONG CASE CLOCK
"This clock can claim to rank with the finest which ever came from the workshops of the former masters."
—*The Watch and Clock Maker.*

T. VICKERY ASTLEY ABBOTTS, BRIDGNORTH, SHROPSHIRE.

15. Long-case Clock made by T. Vickery of Astley Abbotts, Bridgnorth.

16. Watchpaper, Shrewsbury.

17. Long-case Clock made by William Marston of Shrewsbury. Once owned by Daniel Owen, the Welsh novelist. (*Drawn by Richard Philpott*)

Long-case Clock made by John Baddiley of Tong. (*Courtesy of Clive House Museum. Drawn by Richard Philpott*)

19. The Tower, Town Walls, Shrewsbury, which was the workshop of John Massey, watchmaker, in 1816. (*Drawn by Richard Philpott*)

Davies, Henry   Shrewsbury, 1842
'26. May. 1842. Walter, s. of Henry & Eliza Davies. Clockmaker of Claremont St. bapt:' (R).

Davies, John   Shrewsbury, 1789-1797
Attended to the clock and chimes of Old St. Chad's church over the period 1726-1738.
'1726. Jn^{O} Davies for Cleaning ye Clock   3. 3. 0.
1729. By Cash pd John Davis for Mending the Clock   0. 1. 0.
1738. Paid John (Davies) his Salary for ringing & Clock & keeping ye Chimes & Cleaning the Gutters.   2. 14. 0'
(CW, St. Chad's).

Davis, William   Wellington, 1851
William Davis, age 53, Clockmaker, Unmarr: of New Street, born Italy (C, 1851).

Davis, William   Shrewsbury, 1789-1797
Watchmaker, of Wyle Cop. Admitted a Burgess of Shrewsbury 1795. Children by Elizabeth, his wife: Joseph *c.* 1789; Frances *c.* 1793; and William, bapt. 5 July 1797, at St. Julian's. 'At Birmingham, on Thursday last, after a long & painful illness, which she bore with patience & resignation, Mrs. Elizabeth Davies, relict of the late Mr. Wm. Davies, of Wyle Cop' (*SJ*, 17 June 1807).

Davis, William   Shrewsbury, 1863
Clock and watchmaker of Shoplatch. 1863 (D).

Dawson, William   Wellington, 1840-1846
Watch and clockmaker, New St. 1840 (D). Church Street 1842-1846 (D). 'William Dawson, Church Street, age 35, Clockmaker. Elizabeth Dawson, age 34, not born in this county. Louisa Dawson, dau: age 2, born in this county' (C, 1841).

Deakin, William   Dawley, 1856
Clock and watchmaker. Also combined the trade of beer retailer, 1856 (D).

Deaves, Richard   Whitchurch, 1702–1780
Clock and Watchmaker. Son of Edward Deaves. Bapt. 14 February 1702, at Whitchurch (R). Attended to the clock of St. Alkmunds church, Whitchurch for the period 1760–1780.
'1760. To Mr. Deaves a Bill for repairing Clock and Chimes
                                                    2. 0. 0.
1772. Mr. Deaves Bill   1. 14. 0.
1773. Mr. Deaves per bill   1. 18. 0.
1775. 3 May. To Mr. Deaves for attending Clock & Chimes
    ½ Year   0. 15. 0.
To Worke done to D^o wch was Necessary   0. 7. 6.
1777. £2. 2. 0. p bill 7/6   2. 9. 6.
1779. Mr. Deaves by Salary   2. 19. 6.
1780. Richd Deaves in part of bill   1. 15. 0.
    To Mr. Deaves in part of bill   1. 16. 0.
1779. Richd Deaves as per bill looking after Clock &
    Chimes   2. 9. 0.
(CW, St. Alkmunds, Whitchurch). Long-case clock, brass dial, 'Deaves. Whitchurch.'

Del Vecchio & Cetti   Wellington, 1841–1856
Jewellers, cutlers, furniture brokers, New Street. 1856 (D). Goetuno Del Vecchio, age 35, jeweller, Samuel Frankel, age 35, jeweller, New Street, Paul Cetti, age 20, jeweller, New Street (C, 1841).

Del Vecchio & Dotti   Wellington, 1840–1851
Barometer makers, New Street. 1840 (D). Watchmakers, jewellers and furniture brokers, New Street. 1851 (D). Goetuno Del Vecchio (*see* previous entry). Dominic Dotti, age 40, jeweller, New Street (C, 1841).

Doncaster, Edwin   Shrewsbury, n.d.
Long-case clock, brass dial, 8 day, oak case, inscribed 'Edwin Doncaster. Salop'.

Dyxson, Roger   Bridgnorth, 1550
'1550. 8d. to Roger Dyxson for mending the Chymes' (BCR).

Biographical List of Clock and Watchmakers

Eccleshall, Charles   Newport, 1861
'Charles Eccleshall, unmarried, age 18, of Upper Bar. Clock & Watchmaker. born Newport' (C, 1861).

Edge, Griffith   Prees, 1668–1669
1668. 'To Griffith Edge for setting the finger to goe & Also for mending the diall 0. 2. 0.'
1669. 'To Griffith Edge for mending Clock & Chimes 0. 0. 3' (CW, St. Chad's, Prees).

Edmunds, John   Shrewsbury, 1783
John Edmunds, Shrewsbury. an. 1783. Watch. (Baillie.)

Edwards, Edward   Ludlow, c. 1840–1861
Watchmaker. Son of Robert and Harriet Edwards, of Raven Lane. Born c. 1840, at Ludlow. Unmarried and living with parents, 1861 (C, 1861).

Edwards, Robert   Ludlow, 1829–1867
Watchmaker, 21, Raven Lane. Born c. 1802 at Birmingham. Wife, Harriet, born c. 1804 at Ludlow. Children by Harriet, his wife: William c. 1829; Rosa c. 1835; Edward c. 1840; Rowland c. 1842; Robert c. 1846. All four sons became watchmakers at Ludlow. 'Edwards—1st Jan; At Ludlow, aged 63 Mr. Robert Edwards, watchmaker' (*EJ*, 9 January 1867).

Edwards, Robert   Ludlow, c. 1846–1900
Watchmaker. Son of Robert and Harriet Edwards of Raven Lane. Born c. 1846, at Ludlow. Unmarried and living with parents in 1861 (C, 1861). He is returned on the Census of 1871: 'Robert Edwards, sen: married, age 25, Watchmaker of Raven Lane, born Ludlow. Sarah Edwards, wife, age 23, born Abdon, Salop. Robert Edwards, jun; age 2, born Ludlow. Richard. E. George, Lodger. Unmarried, age 19, Watchmaker. born Newtown, Montg:' (C, 1871). He was in business at 90 Lower Gaolford 1879 (D), and at 67 Corve Street, 1891–1900 (D).

Edwards, Rowland   Ludlow, *c*. 1842-1900
Clock and watchmaker. Son of Robert and Harriet Edwards, of Raven Lane. Born *c*. 1842 at Ludlow. Unmarried and living with his parents in 1861 (C, 1861). In business at 7 The High Street, 1868-1885 (D), and at 3 Raven Lane, 1891-1900 (D). The Census Return of 1871 enters him as Rowland Edwards, marr; age 29, watchmaker, High Street, born Ludlow. Mary Edwards, wife, age 28, born Brecon. Harriet Edwards, dau; age 3, born Ludlow. Mary Edwards, dau; age 1, born Ludlow. Edward Watkiss, nephew, age 11, watchmakers apprentice, born Chester (C, 1871).

Edwards, William   Ludlow, *c*. 1829-1851
Watchmaker. Son of Robert Edwards, watchmaker, and Harriet, his wife. Born *c*. 1829 at Ludlow. Living with his parents in 1851 (C, 1851).

Edwards, George   Ludlow, 1849-1850
Watchmaker of Raven Lane, 1849-1850 (D).

Edwards, Edward   Bishopscastle, 1771-1791
Clock and watchmaker, Bishopscastle. 1789 (*UBD*). Edward Edwards, Bishopscastle 1791 (Baillie). Children by Susannah, his wife: Elizabeth 1771; William 1776, who became a watchmaker (R).

Edwards, —   Oakengates, 1868
Watchmaker of Market Street, 1868 (D).

Edwards, Samuel   Dawley, 1856
Watchmaker of Church Lane, 1856 (D).

Edwards, Samuel   Bridgnorth, 1841
Watchmaker of Church Lane. Born *c*. 1791. Children by Mary, his wife: Samuel *c*. 1825; John *c*. 1827; Thomas *c*. 1829; Mary *c*. 1832. His son, Samuel became a watchmaker.

Edwards, Samuel   Bridgnorth, *c*. 1825-1856
Watchmaker. Son of Samuel Edwards, watchmaker, and Mary,

Biographical List of Clock and Watchmakers 57

his wife. Born c. 1825. 'Samuel Edwards, Church Street, aged 25, Unmarried. Journeyman watchmaker, born Bridgnorth.' (C, 1851).
'13.2.1854. at Kidderminster church, by the Rev: S. Peel, Mr. Samuel Edwards, Watchmaker, of High Street, Bridgnorth to Miss. M. A. Shaw, milliner of the same. married.' (EJ). Samuel Edwards. Clock and watchmaker, Bridgnorth, 1856 (D).

Edwards, William Shrewsbury, 1796–1807
Watchmaker, St. Johns Hill. Admitted a Burgess of Shrewsbury, 1796 (PB).

Edwards, William Bishopscastle, 1776–1851
Clock and watchmaker. Son of Edward and Susannah Edwards. Bapt. 7 December 1776 at Bishopscastle (R). Watch movement: 'Wm. Edwards. Bishops Castle.' No face. 'No. 1290. (CHM). Clock and watchmaker, Church Street, 1822/3–1850 (D). Shown on the Census of 1841 and 1851 as unmarried, and on the latter as retired, living with his sister, Elizabeth, also unmarried.

Ellis, George Oakengates, 1842–1850
Watchmaker, 1842–1850 (D).

Ellis, Samuel Wellington, 1863
Watchmaker of New Street, 1863 (D).

Eston, Samuel Ludlow, 1863
Clockmaker of 47 Corve Street, 1863 (D).

Evans, James Shrewsbury, 1709–1774
Watchmaker. Son of Jenkin Evans, Dissenting Minister of Oswestery, and Elizabeth, his wife. Born 13 July 1709, at Oswestry. Admitted a Burgess of Shrewsbury 1732. Charles Barrett bound apprentice to him in 1747, also James Bearley in 1748 (CHR). Churchwarden of St. Julian's 1748. Address given as the Market House in 1768 (PB), and the Cornmarket in 1774 (PB). Children by his wife, Martha: Price James 1746; Richard (n.d.); Martha 1743. Both of his sons became watchmakers.

Evans, Price James   Shrewsbury, 1746-1796
Watchmaker. Son of James Evans, watchmaker, and Martha, his wife. Baptised 25 December 1746 (R). Admitted a Burgess of Shrewsbury 1774. Brother to Richard Evans, watchmaker. Appears on the Burgess Lists 1768-1796. Address given as the Cornmarket.

Evans, Richard   Shrewsbury, 1754-1783
                                                      Oswestry, 1786-1795
Watchmaker and goldsmith. Son of James Evans, watchmaker and Martha, his wife. Bapt. 17 March 1754. As a goldsmith entered his mark as a smallworker 3 May 1779. Admitted to the Company of Mercers, Goldsmiths, etc., of Shrewsbury, 1780, and a Burgess of the town in the same year. Children by Elinor, his wife: John 1783; George 1786; Charles, bur. 1787; Martha 1788 (R). Richard Evans, watchmaker, Oswestry 1789 (*UBD*). Shop in Cross Street. Member, Oswestry General Assoc. for the Prosecution of Felons (1795).

Evans, John   Shrewsbury, 1783-1824
Watchmaker. Son of Richard Evans above. Bapt. 1783 at St. Chad's, Shrewsbury. Admitted a Burgess of Shrewsbury 1806. Address: Cornmarket 1806-1819 (PB). Watch inscribed: 'Jno. Evans. Salop. No: 4009.' Silver case, hall mark: 1814/15 (CHM). Married Mary Donaldson at Holy Cross, 1813. Children: Price James 1814; John 1818; George 1820; Eleanor Ann 1822; and Robert Lloyd 1824 (R). Philip (n.d.). He appears to have moved to Abbey Foregate later on. 'John Evans. Watch and Clock Maker. Market Place. Begs to acquaint the Nobility, Gentry, & the Public in general, he is appointed (Solely) to vend Fribourg and Treyers genuine SNUFF and TOBACCO in Shrewsbury.' (Advert., *SJ*, 29 September 1813.)

Evans, Mary   Shrewsbury, 1848-1862
Widow of John Evans above. Mary Evans, watchmaker, Wyle Cop 1849-1856 (D). 'Mary Evans, widow, aged 55, watchmaker, Wyle Cop. born Shrewsbury. son—Philip, age 27, watchmaker, born Shrewsbury.' (C, 1851). *See also* 1861 Census: 'Mary

Biographical List of Clock and Watchmakers 59

Evans, age 63, widow, retired watchmaker. son—Philip Evans, age 37, Unmarr: watchmaker.' Mary Evans died 29 May 1862.

Evans, Philip Henry Shrewsbury, c. 1823-1862
Watchmaker. Son of William Evans, watchmaker, and Mary his wife. Bapt. c. 1823. Shown on the Census Returns of 1851 and 1861 as a watchmaker of Wyle Cop, unmarried. Admitted a Burgess of Shrewsbury 1844. Philip H. Evans, Watchmaker, Wyle-Cop. 1862 (V.L.).

Evans, George Edward Oswestry, 1868-1885
Watchmaker and jeweller, Cross Street. 1868-1879 (D). Oswald Street. 1885 (D). Watchpaper: 'G. E. Evans. Clock and Watchmaker, Silversmith and Jeweller. Oswestry. Mourning and Wedding Rings. Patent Levers. Jewellery nearly Repaired.' (CHM). 'George R. Evans, Cross Street, marr: age 26, Watchmaker & Jeweller. born Maidstone, Kent. Charlotte Evans, wife, aged 21, born Oswestry. Mary Jane Evans, sister, born Welshpool. Charlotte Evans, dau: age 9 months, born Oswestry. John Pearce, Apprentice Watchmaker, Unmarr: age 17, born Ellesmere (C, 1871).

Evans, John Shrewsbury, 1796
Clockmaker of Hill's Lane. Admitted a Burgess of Shrewsbury, 1796 (PB).

Evans, William Shrewsbury, 1812-1847
Clockmaker. Gullet Shut, 1812 (PB). Coleham, 1814 (PB). Admitted a Burgess of Shrewsbury 1812. Long-case clock, painted dial, 8 day, mahogany banded oak case. Inscribed: 'W. Evans. Shrewsbury.' Children by Mary, his wife: Philip Henry; William; and James; all three sons became clockmakers. 'William Evans, age 50, Wyle-Cop, Watchmaker. Mary Evans, age 50, wife. James Evans, son age 20 (C, 1841). William Evans, died, aged 58, 18 December 1847.

Evans, James Shrewsbury, 1830-1858
Clockmaker. Son of William Evans, watchmaker of Gullet Shut,

and Mary, his wife. Admitted a Burgess of Shrewsbury 1844. Mardol, 1830 (PB). Wyle-Cop, 1847 (PB). 'Lodger at 10, High Street, age 30 (C, 1851). Long-case clock, 8 day, brass dial, 'James Evans. Salop.' Mahogany case. 'Deaths.—EVANS. 9th Feb: aged 37, MR. James Evans, watchmaker, Wyle-Cop, in this town; deeply regretted by a large circle of friends and acquaintainces' (SJ, 17 February 1858).

Evans, Philip Henry   Shrewsbury, c. 1821–1862
Clockmaker. Son of William Evans, watchmaker, and Mary, his wife. Bapt. c. 1821. Brother to James Evans, clockmaker, and to William Evans, watchmaker. Admitted a Burgess of Shrewsbury, 1844. Wyle-Cop, 1844–1847 (PB). Watchmaker, Wyle-Cop, 1862 (VL).

Evans, William   Shrewsbury, 1822–1846
Watchmaker. Son of William Evans, clockmaker of Gullet Shut and Mary, his wife. Brother to James and Philip Henry Evans. Admitted a Burgess of Shrewsbury, 1839. Wyle-Cop, 1822–1846 (D). Wyle-Cop, 1826–1841 (PB). Age 20, on 1841 Census, watchmaker of Coleham Head.

Evans   Shrewsbury
Watch: 'Evans. Shrewsbury.' Enamel face. 'No. 1160' (CHM).

Evans, Richard   Shrewsbury, 1806–1814
Richard Evans, senior, watchmaker, Cornmarket (PB. 1806–1814).

Evans, William   Wellington, 1849–1850
William Evans, New Street, watchmaker. 1849–1850 (D).

Evans & Barnett   Shrewsbury, 1778
Evans & Barnett, Shrewsbury. 1778 (Baillie).

Evans & Brown   Shrewsbury, 1841–1922
Watchmakers. 13 Wyle-Cop. 1863–1922 (D). Partnership of William Evans, son of William and Mary Evans, clockmaker of

Gullet Shut, and Henry Brown, born Newtown, Montg. c. 1821. Watchpaper: 'Evans & Brown. Watch & Clockmakers, Wyle Cop, Shrewsbury. Gold Rings' dated 1907 (CHM). 'Messrs Evans & Brown. Clock and Watchmakers, Jewellers & Opticians, 13, Wyle Cop, Shrewsbury. Established 60 years'. (Advert., *SJ*, 1875).

Evans & Hunt Ellesmere, 1861–1863
Evans & Hunt. Clock and watchmakers, Watergate Street. 1863 (D). 'Robert Hunt, lodger, Charlotte Row, age 59, Watchmaker, born Horncastle, Lincs.' (C, 1861).

Everall, John Minsterley, 1815
'Mr. John Everall, clockmaker, of Minsterley, aged 56, died at Presteigne, Radnorshire' (*SM*, December 1815).

Farmer, Joseph Ludlow, 1851
Watchmaker, Old Street. 1851 (D).

Felton, George Bridgnorth, 1780
Clockmaker. Noted by Baillie and by Britten, 1780. Long-case clock, 8 day, brass dial, oak cased.

Felton, Richard Ludlow, 1711–1732
 Bridgnorth, n.d.
Clockmaker. '24 Julij 1717. Then Ric^d Felton was Enrolled haveing set himself apprentice to $Thom^s$ Vernon to ye trade of Goldsmith, clock & Watchmaker for y^e terme of seven years by Ind bearing date ye first day of may 1711. Richard felton.' '17. Junij 1718. Then $Rich^d$ ffelton was admitted a $freem^n$ to ye tradeof a Goldsmith, Clock & Watchmaker haveing Serv'd an apprenticeship to Mr Thomas Vernon one of y^e $ffreem^n$ of this Company paying to Mr Steward James Wilcox Twenty shillings to y^e use of ye Company. Richard ffelton' (GHR). Married Mary Griffiths, of Ludlow 1732(R). Long-case clock, 30 hour, single finger, brass face, inscribed 'Richard Felton. Bridgnorth.'

Fesser, Andrew Shrewsbury, 1841-1851
Clockmaker. 'Andrew Fesser, aged 25, clockmaker, Frankwell' (with Lawrence Huber, and George Haderer, all clockmakers, all of German origin) (C, 1841). 'John Fesser, a german, Frankwell bur: 3.7.1840.' (R). Andrew Fesser, Clockmaker, Mardol. 1851 (D).

Finn, Thomas Whitchurch, 1856-1863
Dealer in clocks, Green-End Steet, 1856-1863 (D).

Fisher, Ebenezer and Margaret Ellesmere, 1839-1846
Ebenezer Fisher, watch and clockmaker, Cross St. 1840 (D). Ebenezer and Margaret Fisher, watchmakers, Cross St., 1846 (D). 'E. FISHER' on the hour dial of the clock on St. Helen's church, Cockshutt, and the date '1839', possibly the date of some major repair or conversion to a pin-wheel escapement.

Fisher, John Oldbury, 1842-1846
Watchmaker of Freeth Street. 1842-1846 (D).

Fisher, Stephen E. Bridgnorth, 1861
'Stephen. E. Fisher, boarder, East Castle Street, Unmarried, age 23, Watchmaker, born Ellesmere, Salop' (C, 1861).

Fletcher, Charles Shrewsbury, 1724-1739
'1729. By Cash p^d fletcher for altering the Hamer of the Clock. 0. 2. 6.
1736. Chas Fletcher for Mending the Clock. 0. 4. 0.
1739. Paid Mr. Fletcher. 1. 10. 0.'
(CW, St. Chad's.) Children: Charles, buried 23 October 1724, and Mary, buried 13 September 1730 (R).

Fletcher, George Shrewsbury, 1720
Clockmaker. 'George, son of George Fletcher, clockmaker & Jane bap: 31. Jan: 1719/20' (St. Chad's R).

Fletcher, George Coalbrookdale, 1803
Clockmaker. Became an articled clerk at the Coalbrookdale Works in 1803.

Biographical List of Clock and Watchmakers

Ford, Hannah Wellington, 1836
  Watchmaker of Church Street. 1836 (D).

Francis, Edward Maesbury, Oswestry, 1756-1766
Clockmaker. Children by Mary, his wife: Edward 1756; Ann 1757; Richard 1758; Mary 1760; Catherine 1764; and John 1766 (R).

Freeman, Edwin Wem, 1861-1875
Clock and watchmaker, High Street. 1863-1875 (D). 'Edwin Freeman, High Street, Watchmaker & Jeweller, aged 23, born Wem' (living with parents, father—Farmer) (C, 1861).

Freeman, Thomas St. George's, Oakengates, 1863
  Watchmaker of St. George's. 1863 (D).

Furber, Thomas Prees, 1740
'1740. pd. Tho: Furber for mending ye Clock. 0. 1. 0.' (CW, St. Chad's, Prees).

Furber, Thomas Bridgnorth, 1841
'Thomas Furber, aged 30, Watchmaker, St. Mary Street. Born in this county' (C, 1841).

Gardener, John Oswestry, 1744-1765
Clockmaker of Leg Street. Married Catherine Moody 1744. Child by Catherine, his wife: Catherine, buried 1761. His wife died in 1761 (childbirth). John Gardener was buried 28 June 1765 (R).

Gentry, John Shrewsbury, 1870-1875
Watchmaker, St. John's Hill. 1870 (VL); 'John Gentry. Practical Chronometer & Watchmaker (from Benson's, London). Pride Hill.' (Advert., *SJ*, 1871.) 1873. Put in Tender of £7.7.0 for the winding of St. Chad's clock. 'John Gentry, Pride Hill. Practical Watch & Chronometer maker, Gold & Silversmith.' (Advert., *SJ*, 1875.) Watch movement: 'Gentry, Shrewsbury. 6392.' No face (CHM)

Gianna, Lewis   Shrewsbury, 1809-1816
'WARRANTED BAROMETERS. A Farmer or Grazier without a weather Glass, Is just like a Mariner, without a Compass. Lewis Gianna, Barometer and Thermometer Maker, At the Trumpet Inn, Mardol. Begs leave to inform the Public, that he still remains in Shrewsbury and will be happy to supply them with good Barometers and improved Thermometers as any in the Kingdom, and at as reasonable Prices. He needs only mention that the false economy of saving a trifle of Money for an Article of such Value and Utility, subjects many to Losses and Injuries they would afterwards be glad to have avoided.
    Perpendicular Barometers from 10s. 6d. to £2. 2. 0d.
    Wheel Barometers from £2. 2. 0d. to £12. 12. 0d.
    And all other Barometers he makes and warrants.
Barometers delivered without any further expense within 30 miles of Shrewsbury. Old Barometers and Thermometers repaired on the lowest terms.
    Letter post-paid will be attended to. March. 12th. 1809.
    N.B.—Lewis Gianna assures his Friends that no person is authorised by him to travel in his name.' (Advert., *SJ*, 1809).
    'Married. Friday last, at Ludlow, Mr. Lewis Gianna, Barometer & Thermometer maker, of this town, to Miss. L. Moses, of the former place.' (*SJ*, 27 February 1811.)
'19.7.1813. Eleanor Gianna of High Street, aged 33, buried' (R). Two years practically to the day, Lewis Gianna remarried:
'5.7.1815. Lewis Gianna, widower & Elizabeth Evans, widow. by Banns (R). The registers of St. Alkmund's record his burial:
'9.3.1816. Lewis Gianna, High Street, aged 38, buried' (R). The *Salopian Magazine* notes his passing: 'Death. Mr. Lewis Gianna, barometer maker of Shrewsbury: his death is imputed to his having slept in a damp bed.' (*SM.*, March 1816.)

Giles, Richard (Senior)   Wellington, 1832-1836
                                 Shrewsbury, 1840-1877
Watchmaker of Jarrats Lane, 1832. New Street, 1834-36 (D). He had three sons by his wife, Elizabeth: James, *c.* 1824, who died at Evesham 8 March 1867; Henry, *c.* 1830, who became a watchmaker at Oswestry; and Richard, 20 April 1834, who became a watchmaker at Shrewsbury. Richard sen. moved to

Biographical List of Clock and Watchmakers

Shrewsbury by 1840. Richard Giles, watchmaker, Shoplatch 1840-1868 (D). Barker St. 1871-1875 (D). Peacock Passage 1877. Watchpaper: 'Giles, Watchmaker. Opposite the Mermaid Inn, Shoplatch, Shrewsbury. Jewellery repaired' (CHM). The Census Return for 1851 illustrates the background of his family life: 'Richard Giles, aged 48, Watchmaker generally. 46, Shoplatch. born Marylebone, London. Wife—Elizabeth, aged 55, born Wellington, Salop. Daughter—Hannah, aged 19, dressmaker, born Wellington, Salop. Son—Richard, aged 17, watchmaker, born Wellington, Salop.' It would appear that his wife Elizabeth died, and that he remarried, for later in the registers of St. Chad's are to be found: '22.5.1863. Sarah, Ann, d: of Richard & Sarah Giles. Shoplatch. Watchmaker bapt:' and '23.11.1864. Charles, s. of Richard & Sarah Giles. Shoplatch. Watchmaker. bapt:'.

Giles, Richard (junior)   Shrewsbury, 1834-1911
Watchmaker. Son to Richard and Elizabeth Giles, above. Bapt. at Wellington, 20 April 1834. Apprenticed to his father. Was in business at Oswestry, possibly with his elder brother, Henry, in 1856, when he married Mary Ann Henby of Buttington. Children by Mary Ann, his wife: Mary Annie 1859, at 'Crescent Cottage', Murivance; Richard Henley 1861, at Swan Hill; Alfred James at Mardol 1864; Margaret Minnie 1866; Edward Henry 1868; William Albert 1871; Elizabeth Gertrude 1873; Florence Martha 1876, (St. Chad's R). 'Richard Giles, Junior, 16, Mardol, Watchmaker. 1863-1908' (D). Watchpaper: 'R. Giles. Watch & Clock Maker, 16, Mardol, Shrewsbury. English and Foreign Watches and Clocks cleaned and repaired' (CHM). Business taken over by Clement Alcock in 1908, closed end of 1978, as a watchmaker's shop.

Giles, Henry   Oswestry, c. 1830-1883
Watchmaker and Jeweller. Son of Richard Giles, senior and Elizabeth, his wife. Bapt. at Wellington, c. 1830. Brother of Richard Giles, junior, of Shrewsbury. Married Harriet Edwards

of Cefn Blodwel in 1853. Watchmaker and jeweller of Cross St. 1856-1863 (D). Watchpaper: 'H. GILES. Clock & Watchmaker. Oswestry. Plate. Gold Rings.' (CHM.) 'Henry Giles, aged 21, unmarr; Journeyman Watchmaker, Cross St. born Wellington, Salop.' (C, 1851). On the Census Return of 1871 he is shown as 'Henry Giles of Cross St. marr; age 41, Watch & Clockmaker, born Wellington, Salop. Harriet Giles, wife, aged 42, born Llanyblodwel, Salop. Henry. R. Giles, son, aged 16; Adelaide Giles, dau: age 12; Oswald Giles, son, age 10; all born at Oswestry.' Henry Giles died in 1883, aged 53, at his residence, Church St., Oswestry (*SJ*).

Gill, Caleb   Wellington, 1861-1868
'Caleb Gill, Watchmaker's Apprentice. Lodger at 79, High Street, Wellington, aged 17, born Wellington.' (C, 1861.) Watchmaker of Church St. 1868 (D).

Gittins, W.   Shrewsbury, 1786-1806
Barometer-maker of Pride Hill. 1786 (D). 'At the Rev: Mr. Curtis's, Dorrington, Mrs. Gittins, wife of Mr. Gittins, barometer-maker, formerly of this town.' (*SJ*, 12 February 1806.)

Gittos, William   Bridgnorth, 1828-1834
Watchmaker of High Town. 1828 (D). High Street. 1828-1834 (D). Long-case clock, 30-hour (movement only), painted dial, 'William Gittos. Bridgnorth'.

Glase, Edward   Bridgnorth, *c.* 1770-1807
Edward Glase, Bridgnorth, *c.* 1770. Long-case clock (Baillie). Children by Elizabeth, his wife: John 1805; Mary 1807 (R).

Glase, Thomas   Bridgnorth, 1878-1851
Clock and watchmaker. Son of Edward and Sarah Glase. Bapt. 2 December 1787 at Cleobury North (R). Watchmaker and jeweller, High St. (High Town), 1822/3 (D), and Gunsmith 1828-1851 (D). 'Tho. Glase, Bridgnorth. Watch. 1790.' (Britten.) 'Tho. Glase, jeweller, age 50, High St.' (C, 1841.)

'Thomas Glase, Watch & Clockmaker, High St. age 63. born Cleobury North.' (C,1851.) Long-case clock, 30 hour, painted face, 'Thomas Glase. Bridgnorth.' Plain oak case.

Glover, George   Shrewsbury, 1818–1823
'Clock & Watch Making, Wyle-Cop, Shrewsbury. G. GLOVER Begs to inform the Nobility, Gentry, and others, he has opened a Shop in the above line, with an assortment of very superior Gold and Silver Watches and Spring Clocks, viz. Gold Repeaters (plain & musical). Gold and Silver Levers, and plain Watches with an elegant Assortment of Ladies Necklaces, Ear-rings, and Brooches, Musical Boxes, &c. &c. &c.
N.B.—G. Glover begs to say, he has practised for some Years on the best and most difficult Part of the Art: such as Chronometers, Duplex, Vergule, Horizontal, and all the various Escapements, which renders him competent to repair them, without the Delay of sending them to London, having been regularly brought up to the Business. Any Commands in the above line will be respectfully attended to.' (Advert., *SJ*, 30 December 1818).
'G. GLOVER, Watchmaker (from London), Opposite Mr. Blunt's Chemist, Wyle-Cop, Shrewsbury. In returning Thanks to the Nobility, Gentry, & Inhabitants of Shrewsbury, begs Leave to inform them of his having transferred the Business to his Son, John Glover, the Continuance of whose hitherto steady Attention, aided by the thorough knowledge he possesses of his Business, he hopes will render him deserving the Continuance of their Support & Patronage. George Glover begs to intimate his Intention of making his Residence permanent in Shrewsbury, still continuing with his Son in the Business as usual. Shrewsbury. May 24. 1823.' (Advert., *SJ*.)

Glover, John   Shrewsbury, 1823–1834
John Glover (manufacturer from London), Wyle-Cop. 1828–1834 (D). 'John Glover, In taking up, with the Intention of continuing, the late Business of his Father, Mr. George Glover, hopes to meet from his Friends, the Nobility, Gentry, and Inhabitants of Shrewsbury and its Vicinity, a Continuance of their kind Patronage & Support, which it shall be his earnest Endeavour, by an unremitting & respectful Attention, to

merit. Repeating, Horizontal, Duplex, and Lever Watches, Musical Boxes, &c repaired in the most superior & efficient Manner. Shrewsbury. May. 24. 1823. (Advert., *SJ*).

'J. A. GLOVER, (from London). Clock & Watch Manufacturer, Salop. Watches, Clocks, Timepieces, &c. Manufactured to any Price; Warrented of the best quality and workmanship. Watches found troublesome and expensive to the Wearer, or that may have been injured by unskilful Hands, efficiently undertaken and rectified—or exchanged for new. Those having Orders to give, or are desirous of suiting themselves advantageously, may rely on being served with a GOOD ARTICLE, & upon the Best Terms. Works transferred into Gold, Silver, or Gilt Cases, equal to the present Fashion, Gold & Silver Plate, Diamonds, Pearls, Jewellery, and Watches, fairly allowed for in Exchange. All Sorts of Foreign & Repeating Watches, Timekeepers, Chronometers, Patent Levers. MUSICAL SNUFF BOXES &c. Carefully and properly Repaired on the Shortest Notice. Together with the Assortment of New Watches, a Quantity of Good Second-Hand ones remain for Disposal both in Gold, Silver, and Gilt Cases, plain or with 'Seconds'. (The same Terms as with Rest Work, Privileges of Exchange for one Year, with Alterations free). Gold Rings, Seals, Keys, Watch and Timepiece Glasses &c &c. patent & common.' (Advert., *SJ*, 3 January 1827.)

Gorsuch (alias Gossage), Thomas Shrewsbury, 1683–1727
Clock and watchmaker. Son of Thomas and Mary Gorsuch. Bapt. at St. Alkmund's, Shrewsbury, 29 July 1683 (R). Married Catherine Causer, at St. Leonard's, Bridgnorth, 28 August 1703 (R). '1701. Thomas Gosage came in free that yeare as a Clock Maker.' (CBR.) Apprentices: 11 November 1717. William Harley apprenticed to Thomas Gorsuch for 7 years; 2 May 1718. Samuel Brodhurst apprenticed to Thomas Gorsuch for 7 years; 25 May 1725. Thomas Unett apprenticed to Thomas Gorsuch. A fine maker, his skill was noted by the Middletons of Chirk Castle: '1721. Feb: 6th. Pd Mr. Gorsuch, Watchmaker, for mending my Lady's repeating watch. 0. 5. 0.' (CCA.) Children by Catherine, his wife: Sara 1705; Catherine 1707, bur. 1716;

William 1709; Edward 1714; Thomas, bur: 1714; Thomas, buried 1716. His son William became a member of the High Street Church, and by trade a Linnendraper. Admitted a Burgess of Shrewsbury 1747.
'Thomas Gorsuch. Shrewsbury. Silver pair case watch, the cock cut away to show the "pendulum". Movement No. 373. Verge escapement. c. 1710.' (LV.M.) Two watch movements: 'Tho. Gorsuch. Salop.' Enamel faces (CHM). Watch movement: 'Gorsuch. Salop 'metal face, glass on cock (CHM). Watch movements in Oldham Museum, and the Ibert Collection. Thomas Gorsuch buried at St. Chads, Shrewsbury, 30 September, 1727. A very fine maker.

Gottlieb, Andrew William　　　　　　Shrewbury, 1870–1891
Watchmaker, of 31 Wyle-Cop. 1870 (VL). 1879–1885 (D). Gottlieb & Son. 1891 (D). 'Watch repairs of every kind carefully done by A. W. Gottlieb, Who can produce his Indentures of Apprenticeship to the Making of Watches. bottom of Wyle-Cop.' (Advert.: *SJ*, 1879.)

Grant, J.　　　　　　　　　　　　　　　Shrewsbury, 1828
J. Grant. Clock and Watchmaker, of Castlegates. 1828 (D).

Green, George　　　　　　　　　　　Bridgnorth, 1870–1871
Clock and watchmaker. Stonesteps and St. Mary Street. 1863 (D). 74 St. Mary Street. 1868 (D). 'George Green, marr: age 48, Clockmaker. born Manchester. Mary Ann Green, wife, aged 49, born Lancaster. George Green, son, aged 14, born Liverpool. Thomas Green son, age 10, born Bilston, Staffs.' (C, 1851). The Census Return of 1861 gives much the same information with the relative increase in ages.

Green, George & Sons　　　　　　　Bridgnorth, 1870-1871
George Green & Sons, watch and clockmakers, 12 High Street. 1870–1871 (D). 'George Green, Unmarr: age 34, Watchmaker, born Liverpool. of the High Street. Mary Green, mother, aged 62, born Lancaster. Thomas Green, brother, age 30, Unmarr: Watchmaker, born Bilston.' (C, 1871.)

Green, John Shrewsbury, 1686-1688
Watchmaker. Child by his wife, Sara: a daughter, Sara, bapt. 1686 (R, St. Chad's.) Appointed Constable for the Castle Ward Within the Walls, Shrewsbury.1688. (SR.BR).

Green, Walter Ironbridge, 1868
Watchmaker. 1868 (D).

Greenfield, Joseph Ironbridge, 1868-1875
Watchmaker of Waterloo Street. 1868-1875 (D).

Gregory, Richard Bridgnorth, 1861
Watchmaker. 'Richard Gregory, lodger, St. Mary Street, age 49, watchmaker, unmarr: born Frome, Somerset.' (C, 1861.)

Griffiths, William Henry Bishopscastle, 1849-1851
Watchmaker. Market Cross. 1849-1851 (D). 'William. H. Griffiths, Unmarr: age 32, Clock & watchmaker. Market-Cross. born Bishopscastle' (C, 1851.)

Griffiths, William Henry Ludlow, 1856-1875
Clock and watchmaker, jeweller and silversmith, 60 Broad Street. 1856-1875 (D). 'William, Henry, Griffiths. marr: age 41, Broad Street, Watchmaker. born Bishopscastle.' (C, 1861). *See* previous entry for same man.

Grosvenor, John Market-Drayton, 1822-1829
Clock and watchmaker of Shropshire Street 1822/3 (D). Wife died 1829 (*SJ*).

Grosvenor, Robert Market Drayton, 1828-1856
Ellesmere, 1861
Clock and watchmaker. High Street, Market Drayton. 1828-1836 (D). Stafford Street. 1841 (C). Church Street. 1849-1851 (D). In 1851 he was also Registrar of Births and Deaths, and Parish Clerk (D). He was presented at the Drayton Court leet in 1828:
    'Also we present Robert Grosvenor of Drayton in Hales aforesaid, clockmaker, for erecting a Bow window on the side

Biographical List of Clock and Watchmakers 71

of the Common Highway called the Staffordshire Street in Drayton in Hales aforesaid within the jurisdiction of this court whereby the said Highway is much streightened and obstructed to the great annoyance and common nuisance of all his Majeste's liege subjects and do amerce him in the sum of five shillings.' (Court Leet, 1828.)

'Robert Grosvenor, age 48, Stafford Street, Watchmaker, born Market Drayton. Maria Grosvenor, wife, age 50, born Kingswinford, Staffs. Robert Grosvenor, son, aged 11, born Market Drayton. Elizabeth Grosvenor, aged 45, sister, (Independant).' (C, 1841.)

In the Census Return of 1851 he is shown as at Church Street. His son, Robert, became a clockmaker. Robert Grosvenor, sen., moved to Ellesmere by 1861: 'Robert Grosvenor, marr: aged 69, Watchmaker & Finisher. born Market Drayton. Maria. K. Grosvenor, wife, age 70, born Kingswinford.' (C, 1861).

Grosvenor, Robert Edwin Ellesmere, c. 1830-1900
Clock and watchmaker, jeweller. Son of Robert and Maria Grosvenor above. Bapt. c. 1830 at Market Drayton. Watchpaper: (Dated 1862) 'Grosvenor. Clock and Watch Maker. Jeweller &c. Ellesmere. Wedding Rings' (CMH). Swine Market Street 1856 (D). Market Street. 1863-1875 (D). 18 High Street. 1879-1900 (D). Shown on Census 1861 as Market St. 'Robert. E. Grosvenor, 18, High Street, Marr: age 41, Watchmaker. born Market Drayton. Mary Grosvenor, wife, age 41, born Ellesmere. William. A. Pearce. age 14, Watchmaker's apprentice. born Ellesmere' (C, 1871). William A. Pearce later became a Watchmaker with his own business in Church Street in 1879 (D).

Grosvenor, John Cotton, Wem, c. 1783-1861
Clockmaker. Born c. 1783 at Whitchurch, Salop. Mentioned on the Census Returns of 1841, 1851, 1861. Ann Grosvenor, his wife, born c. 1799 at Thorp, Derbyshire. Children: John c. 1830; James 1833; Thomas c. 1835 (became a miller); Robert c. 1834 (became a labourer); Samuel c. 1844. Ann Grosvenor was a laundress in 1861.

Gutteridge, Job Wellington, 1871-1875
Watchmaker of New Street. 1871-1875 (D).

Hall, Robert   **Oswestry, 1800–1838**
Clock and watchmaker. Cross Street. 1822/3–1836 (D). Married 25 February 1800 at Oswestry, Ann Edwards, widow (R). '21.2.1838. On Monday last, Mrs. Hall, wife of Mr. Hall, watchmaker, Cross, Oswestry. Death' (SJ).

Hammonds, John   **Caynham, 1835**
Clockmaker. '26.7.1835. Martha, Sarah, dau: of John & Sarah Hammonds, Clee Hill. Cainham. Clockmaker. born 29. July. bapt:' (LWM).

Hanny, James   **Shrewsbury, 1835–1875**
Clock and watchmaker, Wyle-Cop. 1836–1875 (D).
'J. HANNY, Clock and Watch Maker, Opposite the Lion Inn, Wyle-Cop, Shrewsbury. Begs to inform the Inhabitants of Shrewsbury and its Vicinity, that he has taken to the Business formerly carried on by the late Mr. S. Newham, and hopes, by the strictest Attention to all Orders entrusted to his Care, and the greatest Punctuality, to secure a Share of their Patronage and Support. Clocks, Watches, and Jewellery of every Description cleaned or repaired on the shortest Notice.' (SJ, 1 October 1834.) He is listed on the 1851 Census Return: 'James Hanny, watchmaker, age 41, of 24, Wyle-Cop. (Master employing 2 men) born Bradford, Wilts: Mary Hanny, wife, aged 42, born Beckington, Somerset. William, son, aged 15, Watchmaker's apprentice, born Shrewsbury. Elizabeth, dau: aged 14, Ellen, dau: aged 11, born Shrewsbury. Alice Mary, dau: aged 7, born Knowbury, Salop.'
'J. HANNY, Clock and Watchmaker, opposite the Lion Hotel, Wyle-Cop, Shrewsbury. Gratefully acknowledges the extensive patronage conferred on him by the nobility, gentry, & public generally, during the past nineteen years, and respectfully solicits a continuation of those favours which it will be his constant study to execute with care, combined with honest and conscientious charges. Clocks and Watches of every description made or obtained to order, cleaned, and repaired. Church and Turret Clocks kept in good state at per annum. Accordions and Concertinas tuned. Gold Wedding Rings.' (SJ, 4 January 1854.) Watchpaper: 'J. Hanny, Clock and Watchmaker,

opposite the Lion Hotel Wyle Cop, Shrewsbury. Wedding Rings. English & Foreign Clocks and Watches of every Description cleaned & repaired.' Dated on back: '9.11.81' (CHM). Watchpaper, similar to above, but handwritten on back: '24.11.75. Mr. Smout, Stiperstones' (CHM). His wife, Mary died aged 67 in 1875. He put in a tender in 1873 to wind St. Chad's clock.

Hanny, William Stourton   Shrewsbury, c. 1836-1865
Clock and watchmaker, Castle-Gates. 1836 (D). Son of James & Mary Hanny, bapt. c. 1836, at Shrewsbury. Apprenticed to his father, James Hanny, of Wyle-Cop. Hanny—3rd June, at Castle-Gates, in this town, aged 1 year & 6 months, Bertha, Ellen, youngest daughter of W. Stourton Hanny.' (*SJ*, 7 June 1865).

Harley, George   Shrewsbury, 1784
George Harley, Shrewsbury. 1784. Watch. (Baillie.)

Harley, William   Shrewsbury, 1717-1764
Clock and Watchmaker. Son of Richard Harley of Shrewsbury, Gent. 'William Harley hath Put him Self An Apprentice to Thomas Gorsuch Clockmaker in Shrewsbury for Seaven Yeares his time begins November the 11th 1717. Rec^d 0-2-6.' (CBR). As a master, he in turn took William Davis as an apprentice c. 1731 (CBR). Later, c. 1749: 'John Dickin hath put himself Aprentis to William Harley Watchmaker for the term of Seven Years Bearing the Date of his indenter Rec^d for the Use of the Company 5s.' (CBR). He is shown on the Burgess List of 1734 as a watchmaker, of the Cornmarket. In 1747 on a similar document he is additionally described as an 'Anabaptist'. In 1740 his daughter, Deborah died, and in 1745 he lost his wife, Hannah (R). Admitted a Burgess of Shrewsbury 1753. His own death is recorded in 1764: '25. July, 1764. William Harley, watchmaker buried' (St. Chad's R).

Harley Samuel   Shrewsbury, 1766-1799
Clock and watchmaker, goldsmith. Son of William Harley, watchmaker and Hannah his wife. Children by Anne, his wife:

William 1766; Anne 1767, bur. 1767; Anne 1768; Samuel 1769; Hannah 1772, bur. 1776; Mary 1777. Admitted a Burgess of Shrewsbury 1767. Samuel Harley, watchmaker, Murivance. 1774 (BL). Later of the Market-Place 1786 (D). Mayor of Shrewsbury in 1784, when he is described as a watchmaker and goldsmith. Long-case clock, brass face, inscribed: 'Saml. Harley. Salop.' Churchwarden of St. Chad's 1785. 'Supplied new Time-Piece to St. Chads. 3. 3. 0.' (CW, St. Chad's). Trustee of land belonging to the High Street Chapel 1778. His son, Samuel became a grocer, in this town, and was admitted a Burgess of Shrewsbury, 1795. His wife, Anne, died, aged 62, in 1799.

Harley, William   **Shrewsbury, 1766-1843**
Watchmaker and goldsmith. Son of Samuel Harley, watchmaker and Anne, his wife. Bapt. 9 January 1766, at Shrewsbury. Married in 1789, Ann Lloyd. Children by Ann, his wife: John 1792; Maria 1795; Elizabeth 1797; William Lloyd 1798; Edward Daker 1801; Samuel, n.d. None of his sons continued in his business. In 1796 he is listed as a watchmaker of Murivance (PB). In 1807 he was resident at Swan Hill (PB). Admitted a Burgess of Shrewsbury 1789, and like his father, became mayor of Shrewsbury. 'William Harley, Watchmaker & Goldsmith. Mayor 1814.' He died in 1843: '6.11.1843. William Harley, late Mayor of Shrewsbury, aged 77, of Bridge Place. buried.' (St. Chad's R).

Harley & Son   **Shrewsbury, 1789-1809**
Harley & Son, Watchmakers and goldsmiths. 1789 (UBD). Long-case clock, brass and silvered dial, 8 day. Also watch movement: 'Harley & Son. Shrewsbury. No. 4358.' Enamel face. (CHM). 'Watch Lost. On Saturday, the 7th instant, near the Butter Cross in the Town of Shrewsbury. A Watch in a Tortoise-shell Case, with the Owner's Name on the Face. Whoever has found it, and will bring it to Messrs. Harley & Son, Watchmakers, Shrewsbury, shall recieve One Guinea Reward. N.B. No greater Reward will be offered' (*SJ* Advert. 1806). 'Harley and Son, Watchmakers, Goldsmiths, and Jewellers, Shrewsbury. Inform their Friends and the Public, that they have just laid in a new and fashionable Variety of Articles in

the Gold, Silver, Jewellery and Watch Trades:—also of Plated Goods, and Cutlery, from the first Manufactories. Fine Teas, Coffee, Chocolate, Glass, China, &c &c. they beg to offer their Friends as usual. Harley & Son having frequently found much Inconvenience from not being able to sell Plate by Auction. W.H. begs leave to say, that on renewing their Licence for the Sale of Plate this year, he has obtained the necessary Authority to sell by Auction.' (Advt.: *SJ*, 13 September 1809.)

Harper, Thomas   **Shrewsbury, 1723-1774**
Watchmaker. Married Catherine Cleveley, both of St. Alkmund's parish, at Holy Cross, 23 May, 1723. Churchwarden of St. Chad's. 1747. Repaired St. Chad's clock in 1750: '1750. 26. Jan. paid Mr. Harper for Work Don at the Clock. 0. 12. 0.' (CW). He took an apprentice in 1723: 'William Harris son of Richard Harris hath with the Consent of his father put himself aprentis to Thos. Harper Clockmaker. His time to begin the 2nd day of Feby 1723/4.' (CBR). Later in 1748, his son, Richard, was apprenticed to him: '1748. Richard Harper son of Thos. Harper hath putt him Self an Aprentis To his ffather for The Term of Seven Years. Recev^d for the Use of the Company 5s.' (CBR). He was elected one of the two wardens of the Smith's Company, of Shrewsbury, in 1748. (CBR). He is shown as a watchmaker of Mardol, in 1768. (BL), and later in 1774 in Castle-Foregate (BL).

Harper, Richard   **Shrewsbury, 1748-1791**
Watchmaker. Son of Thomas Harper, Watchmaker, and Catherine his wife. Apprenticed to his father in 1748 (*see* previous entry). Both Baillie and Britten speak of him as an eminent watchmaker. He was one of the original 30 proprietors of the Salop Fire Office. His death is noted in the *Shrewsbury Chronicle* under the date, 11 February 1791: 'On Monday morning last died, at his house on St. John's Hill, Mr. Richard Harper, formerly an eminent watch-maker in this town, but who had retired from business several years.—He was respected, by a numerous acquaintance, as a sincere and valuable friend.' His widow, Sarah, lived on to the great age of 83, dying in 1815.

Harper & Son Shrewsbury, 1775
'Harper & Son. Salop. Clock about 1775.' (Britten.)

Harper, – Shrewsbury, c. 1790
'Harper, –, Shrewsbury, ca. 1790.' (CC). 'In firm of Carswell & Harper.' (Baillie.)

Harris, William Shrewsbury, 1723/4-1736
Watchmaker. Son of Richard Harris, Innkeeper of Shrewsbury. He was apprenticed to Thomas Harper in the February of 1723/4. He married in 1736, Mary Tomkies, of Shrewsbury. Long-case clock, 8 day, brass face, 'William Harris. Salop.' Mahogany case.

Harris, Richard Shrewsbury, 1796
Watchmaker, of Hill's Lane. 1796. (PB). Admitted a Burgess of Shrewsbury 1796 (PB). Children: Elizabeth, c. 1769; Richard c. 1774; William c. 1775; John c. 1777; Mary c. 1779. Sarah c. 1789; Martha c. 1792 (PB).

Harris, Richard Wellington, 1738-1765
Clock and watchmaker. Son of Richard and Isabell Harris. Bapt. 12 November 1738 at Wellington (R). Married Elizabeth Harrison on 22 August 1765 at Wellington (R). Long-case clock, 'Richard Harris. Wellington'. 8 day, brass and silvered dial, oak and mahogany banded case.

Harris, William Bridgnorth, c. 1780
'William Harris, Bridgnorth. ca. 1780. Clock.' (Baillie.)

Harrison, John Market Drayton, 1822-1836
Clock and watchmaker of Cheshire Street. 1822/3-1836.

Hartshorne, William Broseley, 1793
Clock and watchmaker. Long-case clock, 36 hour, square-faced painted dial inscribed: 'W. Hartshorne. Broseley' oak-case. 'William Hargshorne, Broseley. 1793. Watch.' (Baillie.)

Harvey, John Wellington, 1861-1879
Clock and Watchmaker, Mount Terrace, 1863-1875 (D). Bridge Road 1879 (D). 'John Harvey, married, age 34, Clock and Watchmaker (Master,) of Mount Terrace, born Orkney Isles. Janet Harvey, wife, aged 34. Janet Harvey, dau: age 5, born Scotland. Henrietta Harvey, dau: age 3, born Wellington. George Harvey, son, age 1, born Wellington.' (C, 1861.)

Haslip, William Wenlock, 1793
Watchmaker. 'William Haslip, Wenlock. an. 1793. Watch.' (Baillie.)

Hasty and Son Shrewsbury, 1784
Watchmakers. 'Hasty & Son, Shrewsbury. an. 1784. Watch.' (Baillie.)

Hawkes, John Forrester Albrighton, 1802-1809
Clock and watchmaker. Married Elizabeth, dau. of Charles Blakeway, clockmaker of Albrighton, in 1802, when he is described as from the parish of Walsall. A triple baptism occurs at Albrighton, in 1809, of his three daughters: Elizabeth Blakeway, Frances Priscilla, and Eleanor (R). 'J. Hawkes. Albrighton. 1810. C & W.' (Baillie.)

Hay, Thomas Bishops Castle, 1787-1801
 Shrewsbury, 1801-1829
Watchmaker. 'Thomas Hay. Bishops Castle. *ca.* 1770. Watch in Shrewsbury Museum' (Baillie.). Married Susannah Woodall at Bishopscastle in 1787 (R). Eight children born at Bishops-Castle during the period 1788-1801, and four buried. He removed to Shrewsbury in 1801, where a further three children were born, and one buried: 1803-1807 (R, St. Alkmund's). His premises were in the High Street, Shrewsbury, 1807. 'On Friday last, at Ryton, Mr. Thomas Hay, of this town, watchmaker, in his 73rd year' (*SJ*, 15 July 1829).

Hay, Thomas William Shrewsbury, 1822-1856
Clock and watchmaker, Market Square, 1822/3-1846 (D). 8 High St., 1849-1854 (D). 7 Market Square, 1854 onwards

(D). 'REMOVAL. Thomas William Hay, Watch & Clockmaker. From 8, High St to 7, Market-Place.' (Advert.: *SJ*, 25 January 1854). Watch movement: 'Thos. Willm. Hay. Shrewsbury. No: 21432.' No face (CHM). Watch movement: similar inscription, enamel face, 'No: 18892' (CHM). His obituary reads: 'Hay—21st May. Mr. Thomas William Hay, watchmaker, of this town, aged 66' (*SJ*, 28 May 1856).

Hay, Cecilia (Mrs.)   Shrewsbury, 1856–1875
Clock and watchmaker. (Widow of above.) Market Square, 1863–1875 (D).
'WATCHES Mrs. Hay Begs to solicit an inspection of the large assortment of Ladies & Gentleman's Gold & Silver Watches, Chains etc with which her son has just returned from London. Selected from the most celebrated English and Foreign houses. Market Square, Shrewsbury' (*SJ*, 6 August 1857). 'HAY. CLOCK AND WATCH MAKER. MARKET SQUARE, SHREWSBURY. This business established in the last century, has been carried on in the above name since 1801, by experienced and able workmen, and by connection with the first manufacturers both in London and Geneva. A large stock is kept of English and foreign Watches, Clocks, and Timepieces, Gold and Silver Guards & Chains, Keys, Seals, Wedding Rings etc, Jewellery etc Repaired–Engraving' (*SJ*, 5 May 1858). 'Cecilia Hay, aged 63, Holywell Lane, Watchmaker & Jeweller, born Middx. London. Francis Hay, dau: aged 20, born Shrewsbury. Susan Elizabeth Hay, aged 20. born Shrewsbury' (C, 1861).

Hay, Thomas William   Shrewsbury, 1857–1862
Watchmaker, son of Tho. Will. and Cecilia Hay. In business with his mother. Market Square. Children by Catherine Martha Llewellyn, his wife: Tho. Will. Llewellyn 1859; Geo. Arthur, Dorsett 1860; Henry James 1862.

Hays, Thomas   Shrewsbury, 1873–1875
Watchmaker, Benbow-Place, Coton Hill 1875 (D). Put in Tender for winding St. Chad's clock 1873. £7 7s. 0d.

Biographical List of Clock and Watchmakers

Hayward, Thomas **Bridgnorth, 1841–1861**
Clock and watchmaker. 'Thomas Hayward, age 50, Clockmaker, Newtown. Elizabeth Hayward, age 50, wife. John Hayward, age 25, son, Carpet Weaver. William Hayward, aged 20, son, Carpet Weaver. Emma Hayward, age 14, dau: James Hayward, aged 12, son' (C, 1841). 'Thomas Hayward, age 63, Watchmaker, Friars Street, born Birmingham. Elizabeth, wife, aged 63, born Shenstone, Staffs. James, son, aged 22, Carpet Weaver, born Bridgnorth' (C, 1851). 'Thomas Hayward, Clockmaker, aged 72, Union Workhouse. born Birmingham' (C, 1861).

Hazeldine, William **Rowton, High Ercall, 1672–1726**
Clockmaker. Made a new clock for St. Mary's, Shawbury, in 1672 (CS). He also made a new clock for St. Chad's, Prees, in 1684: 'To Hazeldine for makeing a new clock and placeing it for repairing ye Chimes and spent upon him 3.13.0.' (CW Accts). William Hazeldine made an agreement to maintain the clock of St. Luke's, Hodnet, in the February of 1691/2: 'The Condicon of this obligacon is Such That if he above bounden Wm Hasledine & his heirs Exec^S & Admins do & shall from time to time & at all times hereafter for & during unto ye full end & terme of twenty yeers next ensuing after ye date of these presents fully to be complete & ended Keep & maintaine ye Church clocke in he parish of Hodnett in ye said County of Salop in good & Sufficient order & repaire at & under ye yeerly Salary of five shillings a yeere to ye abovebounden Wm Hasledine by ye Churchwardens of ye said parish of Hodnett to be well & truly paid yeerly, during all ye said Terme & Space of twenty yeers according to ye true intent & meaning of a certaine agreement made between Charlton Hill & Rich^d mullinex Churchwardens of ye said parish of Hodnett & ye abovebounden Willm Hasledine & entred into ye parish booke of ye said parish & bearing equal date with p'sents. That then this present obligation be void & of none effect or els ye same to Stand & be full force effect & vertue. William hasildene. Sealed deliver'd in p'sence of John Rowley. Nat. Cureton, Cur^t.' (SRO. 2275/70A). In 1698, the churchwardens of St. Chad's,

Stockton, paid: 'Wm Hazeldine the clockman 8. 8. 0.' (CW). He married Anne Buttery of Rowton, 25 November 1782. Children by Anne, his wife: Anne 1684; William 1687; Mary 1689; Margaret 1691; Sarah 1693; Richard 1696, buried 1712; Thomas 1703; and Elizabeth 1706. About 1693 he appears to have moved from Rowton to Cold Hatton. He lived to the fine old age of 80, having been born in 1646, and died in 1726:
'1726. Feb: William Hazeldine, of Cold Hatton buried' (R).

Hazeldine, - -, 1753
It would seem that one of the two surviving sons of William Hazeldine—William 1686, or Thomas 1703— must have carried on his father's business after his death, for in the churchwarden's accounts of All Saints, Berrington, we find the following entry: '1753. Pd Hazeldine for work at the Clock. 0. 18. 0.'

Henshall, Henry Whitchurch, 1863
Watchmaker of Paradise Street, 1863 (D).

Herbert, William Ludlow, 1820-1842
Clock, watchmaker and Jeweller. High Street, 1822/3-1836 (D). Admitted to the Ludlow Guild of Hammermen 26 December 1820 (GHR). Long-case clock, 30 hour, painted dial. 'W. Herbert. Ludlow', oak case with mahogany banding. 'On the 31st ult. suddenly, Mrs. Herbert, wife of Mr. William Herbert, watchmaker, Ludlow' (*SN* & *CR,* 9 April 1842).

Herbert & Son Ludlow, 1840-1850
Watch and clockmaker, and silversmiths. Broad Street. 1840-1850 (D). A partnership of William Herbert, senior, and his son, Thomas Herbert.

Herbert, Thomas Ludlow, 1851
'Thomas Herbert, Broad Street, Unmarr: age 36, Watchmaker, employing one man. born Ludlow' (C, 1851).

Higgons, Joseph Bridgnorth, 1680
In 1680 the corporate body set up a new clock on the Gild (or Town) Hall, with a dial-plate at each end, which was made by Joseph Higgons at a charge of £8.

Higgs, Thomas Ludlow, 1565–1606
The Churchwarden's accounts of St. Laurence, Ludlow, show this man to have attended the church clock and chimes for the greater part of the reign of Elizabeth:
  '1565–6. Item to Thomas Higges, for Keping the clocke and chymes   vjs. viijd.
  1593–4. Item to Thoms Higgs for Keping the Chymes for ye hole yere   vjs. viijd.
  1599–1600. Item paied to Thomas higgs for Keepinge the Clock and Chimes   vjs. viijd.' (CW).
  '15.8.1606. Thomas Higs, deacon buried. He was of an hundred years & had been deacon 50 years' (R).

Higgs, William Oswestry, 1778
'William Higgs, Oswestry. 1788' (Baillie).

Highfield, William Oswestry, 1775–1782
Watchmaker of Bailey Street. Baillie puts him as of Liverpool in 1761, and later at Wrexham, then at Oswestry, 1778. The parish registers of St. Oswald's, Oswestry, give the following information: burial of his son, Nathaniel, 27 February 1775; baptism of his son, John, 15 July 1775, and his own burial:
  '15.9.1782. William Highfield, Bailey Street. bur:' (R).

Hill, Thomas Wem, 1841–1851
Watchmaker, Noble Street, 1842–1846. High Street, 1849–1850. Market Street, 1851 (D). 'Thomas Hill, watchmaker, aged 45, Noble Street. Ann Hill, aged 45. George Hill, aged 10. Elizabeth Hill, age 7' (C, 1841). 'Thomas Hill, marr: age 59, Watchmaker (Master) High Street. born Prees. Ann Hill, wife, aged 59, born Malpas. Elizabeth Hill, daughter, age 19, born Malpas. Elizabeth Hill, grand-daughter, age 5, born Wem.' (C, 1851).

Hilton, Evan Shrewsbury, 1667
Watchmaker of St. Chad's parish. Son, William, baptised 29 March 1667 (R, St. Chad's).

Hinckley, William Ironbridge, 1849-1851
Watchmaker, of Bridge Street. 1849-1851 (D). 'William Hinckley, marr: age 29, Watchmaker, born Birmingham. Elizabeth Hinckley, wife, age 30, born Bewdley, Worcs. Mary Hinckley, daughter, age 3, born Ironbridge' (C, 1851).

Hinksman, James Sutton Maddock, 1760
James Hinksman, Sutton Maddock. 1760. Clock' (Baillie).

Hinksman, James Bridgnorth, 1789
'James Hinksman. Bridgnorth. 1786-94. Watch' (Baillie). 'James Hinksman. Clockmaker. 1789' (UBD). Long-case clock, painted dial. Long-case clock, 30 hour, heavily engraved and silvered dial.

Holmes, Samuel Wellington, 1828-1834
Watchmaker, of New Street. 1828-1834 (D).

Hopwood, Robert Bridgnorth, 1861-1875
Watchmaker. 8 Waterloo Terrace. 1863-1868 (D). Jeweller, silversmith, watchmaker, and engraver. 24 High Street. 1870-1875 (D). 'Robert Hopwood, Unmarr: age 24, Watchmaker & Engraver, Waterloo Terrace. born Cirencester, Clouc: Jane Hopwood, sister, Unmarr: age 28, born Cirencester, Glouc:' (C, 1861). 'Robert Hopwood, marr: age 34, Watchmaker & Jeweller, employing one man, High Street. Born Cirencester, Glouc: Harriet Hopwood, wife, age 33, Photographer, employing one man. born Greenwich. Ethel Hopwood, daughter age 5, born Bridgnorth. Harold. M. Hopwood, son, age 3, born Bridgnorth' (C, 1871).

Houghton, Richard   **Prees, 1670-1671**
'1670. paid to Richard houghton for mending the Clocke and Chimes 4. 15. 4.
1671 Pair to Richd Horton & his Sonn for getting the Clock & Chimes in order 2.10.0.
Paid William Pye ffor the us of his Shop for Richd Horton for Iron and Coles towards repaireing of clocke and Chimes 2. 0. 0.' (Prees, CW).
N.B.—William Pye was the local blacksmith.

Huber & Co.   **Shrewsbury, 1841-1846**
Clockmakers of Frankwell, 1842-1846 (D). 'Lawrence Huber, a german, Frankwell, bur: 15.9.1842' (R, St. Chad's). 'Lawrence Huber, age 25, Clockmaker, Frankwell' (C, 1841). At the same address as Andrew Fesser, age 25, and George Haderer, age 15, all clockmakers.

Huber and Fesson   **Wellington, 1842-1846**
As above at Shrewsbury. German clockmakers, Watling Street, 1842-1846 (D).

Hugh Ap William   **Oswestry, c. 1590**
c. 1590. 'Item pd to hughe ap Wm for mendinge the clocke Vd.' (CW, St. Oswalds).

Hulse, Ralph   **Market Drayton, 1789-1791**
Clock and watchmaker. 1789 (UBD). Ralph Hulse. Market Drayton. 1791. C & W. (Baillie.)

Hunt, Robert   **Ellesmere, 1861-1863**
Watchmaker. Partner in Evans & Hunt. Clock and watchmakers, Watergate Street. 1863 (D). 'Robert Hunt, lodger, Charlotte Row. age 59, Watchmaker. born Horncastle, Lincs' (C, 1861).

Hutton, —   **Shrewsbury, 1751**
'1751. 21. April. p^d Hutton for mending ye Clock. 0. 3. 6.' (CW, St. Chad's).

Ireland, John   **Shrewsbury, c. 1760**
'John Ireland. Shrewsbury. *ca.* 1760. Watch movement. Shrewsbury Museum' (Baillie).

Jackson, William   **Shrewsbury, 1827**
'1827. Jan. 18. Charles, son of William & Margaret Jackson (House of Industry) watchmaker. bapt:' (R, St. Chad's).

James, David   **Market Drayton, 1861**
'David James, lodger, Church Street. Unmarr: age 20, Watchmaker. born Brecon' (C, 1861).

Jared, Ann   **Selattyn, 1744-1786**
Clockmaker. Children by her husband, William: William 1744; Ann 1745; Esther 1746, bur 1761; Ann 1750; Edward 1752; James 1754, bur. 1756; George 1758; and Hester 1762. Her husband, William Jared died, aged 74, and was buried 28 August 1786. Possibly also a clockmaker, but no evidence. Two of their sons (George and William) became clockmakers. 'Ann Jared, widow, clockmaker, Pantglas, aged 72, buried 28.8.1786' (R).

Jared, George   **Oswestry, 1758-1788**
Clockmaker, of Willow Street. Son of William and Ann Jared of Selattyn, bapt. there on 28 February 1758. Wrongly noted by Baillie as 1791, also listed in the *Universal British Directory*, published in 1790, although he died in 1788. Children by Elizabeth, his wife: George 1784; and Hannah 1785 (R). 'George Jarrett, clockmaker, Oswestry. died 17.12.1788. aged 31. (decay & fever) buried 20. Dec: 1788' (R).

Jared, William   **Selattyn, 1744-1786**
Clockmaker of Selattyn. Son of William and Ann Jared. Bapt. 27 May 1744 at Selattyn(R). Married Mary Noden, at Ellesmere, 19 November 1770. A daughter, Elinor, was born 20 May 1779, died aged 27, in 1807. His wife, Mary, died aged 56, on 19 May 1798. 'William Jarred, clockmaker. buried 31. Jan: 1804, aged 60. (Gout)' (R).

Biographical List of Clock and Watchmakers 85

Jarvis, John   Whitchurch, 1842-1879
Clock and watchmaker, 7 Green End. 1846-1879 (D). Children by Mary Ann, his wife (all born at Whitchurch): Elizabeth, c. 1847, Harriet, c. 1849; James c. 1854. and Mary Ann c. 1856. Watchpaper: 'John Jarvis. Watch and clockmaker. Whitchurch, Patent Lever Watches. All kinds of Clocks, Watches, Musical Boxes, Plate, Jewellery, Cleaned & Repaired' (CHM). Married Anne Baxter: 'On the 15th inst. Mr. John Jarvis, clock & watchmaker, to Miss. Anne Baxter, both of Whitchurch' (*SN & CR*, 26 November 1842). 'John Jarvis, watchmaker, Green End, age 50, born Whitchurch, Salop. Mary Ann Jarvis, wife, age 58, born Ashby de la Zouch. Elizabeth Jarvis, dau: age 24, born Whitchurch. Harriet Jarvis, dau: age 22, born Whitchurch. James. S. Jarvis, son, age 17, watchmaker. born Whitchurch. Mary Ann Jarvis, dau: aged 15, born Whitchurch' (C, 1871).

Jeffries, Edward   Much Wenlock, 1778
Clock and watchmaker. 'Edward Jeffries of Much Wenlock, clock & watchmaker, & Mary Russell, spinster. married by lic: 15. March. 1778'. (St. Bartholomews Registers, Benthall).

Jepson, Forrester   Shrewsbury, 1868
                                                  Bishopscastle, 1868
Watchmaker of 30 Mardol, Shrewsbury, and at Church Street, Bishopscastle. 1868. (D).

Jervis, Henry   Newport, 1730
Clockmaker, of Newport. Agreement in Vestry minutes of St. Swithun's, Cheswardine regarding the upkeep of the church clock: 'Agreed with Henry Jervis of Newport, Clockmaker to put the Church Clock of Cheswardine into good repair & so to keep it till May day 1731 from this time (May 1730) & to turn the weights of ye Clock to come down within ye rails in the corner of ye Steeple where formerly they did & the churchwardens when finished to pay him fifteen shillings & after May day 1731 to have every year 2s. 6d. for keeping the Clock in repaires all that shall want by the wearing of it or bushing or pulleyes. Accidental chances and bruises or damage & ropes excepted, by H. Jervis in this bargain' (VM, Cheswardine).

Jones, Henry Charles   Shrewsbury, c. 1808-1838
Watchmaker, Claremont Street. Admitted a Burgess of Shrewsbury 1830 (PB). Son of John Jones, printer of Dog Lane. Children by Mary, his wife: William Frederick 1834; Edward 1832; Edwin Rowland 1836; and Alfred 1837 (R). '12.2.1838. Henry Charles Jones, watchmaker of Barker Street buried, aged 30' (St. Chad's R).

Jones, James   Shrewsbury, 1861-1865
Watchmaker. The 1861 Census returns him as a lodger, unmarried, aged 20, on Pride Hill. '31.8.1865. Edith, daughter of James & Ann Jones. St. Austins St. Watchmaker. bapt:' (St. Chad's).

Jones, Joseph   Oswestry, 1861-1885
Clock and watchmaker, Bailey Head. 1863 (D). Willow St. 1868-1885 (D). Watchpaper: 'Joseph Jones. Watch and Clockmaker, Bailey Head. Oswestry' (CHM). 'Joseph Jones, Bailey Head. Marr: age 45, Watchmaker, born Berriew, Montg; Mary Jones, wife, age 44, born Hirnant, Montg: Sarah Jones, dau: age 7, scholar, born Oswestry' (C, 1861). N.B.—Discrepancies in ages are commonplace as given by people on Census Returns; with this small family, Sarah ages correctly 10 years on the 1871 Census, but Joseph Jones leaps forward 15 years, and his wife, Mary, 17 years.

Jones, Humphrey   Oswestry, 1801-1848
Clock and watchmaker, Bailey St. 1822/23-1846 (D). Children by Elizabeth, his wife: John 1801; Elizabeth 1802; Mary, bur. 1804; Thomas 1805; William 1808, bur. 1812; Jane 1809; Catherine 1810, bur. 1811; and Ann 1812. His wife, Elizabeth, died 1820. He married secondly, '5.11.1821. Mr. Humphrey Jones, watchmaker, to Miss Mary Mitchell, both of Oswestry' (*SJ*, 7 November 1821). The marriage was not fated to last long, for she died on 6 November 1824. 'On the 4th inst. after a lingering illness, in his 76th year, Mr. Humphrey Jones, watchmaker, of Oswestry' (*SJ*, 8 November 1848).

Biographical List of Clock and Watchmakers

Jones, Thomas Oswestry, 1832
'On the 20th inst. after a lingering illness, Mr. Thomas Jones, watchmaker, Oswestry.' (*SJ*, 29 September 1832).

Jones, William Ludlow, 1822/3-1834
Watchmaker, of the High Street. 1822/3-1834 (D).

Joyce, William Cockshutt, 1692-1771
Clockmaker. Son of John and Elizabeth Joyce, of the Lodge, Cockshutt. Bapt. at Ellesmere, 9 February 1691/2 (R). First appears as a clockmaker at Wrexham, where two of his children are baptised: John 1718; and Elizabeth 1721. He was in all probability apprentice to John William Joyce, clockmaker of Wrexham, who died in 1717, and most probably took over his business. William Joyce was paid £2 8s. 0d. for attending to the clock and chimes of Wrexham parish church in 1718. He moved back to his birthplace, Cockshutt, by 1723, when his son, Arthur, was baptised at Ellesmere. His wife, Mary, was buried 1724. He then married Ann Jones at Ellesmere in 1725. Buried 17 February 1771.

Joyce, John Ellesmere, 1718-1787
Clockmaker. Son of William and Mary Joyce, bapt. at Wrexham, 24 May 1718. Married Deborah Sadler, at Ellesmere, 1738. Attended the clock of St. Mary's, Ellesmere, for 20 years— 1738-1758. Children by Deborah, his wife: Elizabeth 1742; John 1744; William 1748; James 1752; Robert 1754; Malachi 1756, bur. 1757; Samuel 1759; and Conway 1761. Of his seven sons, five became clockmakers. His wife, Deborah was buried at Ruthin in 1799 while John Joyce was buried at Cockshutt in 1787.

Joyce, James Whitchurch, 1752-1817
Clock and watchmaker, High Street. Son of John and Deborah Joyce. Bapt. at Ellesmere, 10 July 1752. Attended to the Clock of St. Alkmund's, Whitchurch, from 1784 until at least 1806. He moved the family business from Cockshutt to Whitchurch sometime prior to his father's death, for he is recorded of Whitchurch, when he married Sarah Barnett, at Wem, 12 November 1782. Children by Sarah, his wife: Elizabeth 1783; Ann

1784; James 1786, bur. 1791; Mary 1787, bur. 1788; Sarah 1791; William, bur. 1791; John Barnett 1792, bur. 1822; Thomas 1793; and Richard Owen 1796. Of his five sons, only Thomas carried on in the family business. Sarah, wife of James, died in 1808, while James Joyce died 23 December 1817.

Joyce, Thomas   Whitchurch, 1793-1861
Clock and watchmaker. Son of James and Sarah Joyce. Bapt. at Whitchurch, 13 October 1793 (R). Married Charlotte Jones, of Llay Hall, Gresford. Watchmaker and engraver, High Street, 1822/3-1846 (D). Watchmaker, maltster and agent to the Provident Life Assurance Office, High Street. 1856 (D). Children by Charlotte, his wife: Charlotte; James 1821; Conway; Emma; Thomas; John Barnett, 1826. Of the four sons, two became clockmakers—James and John Barnett. Thomas Joyce died 25 February 1861.

Joyce, Thomas & Son   Whitchurch, 1849-1851
Partnership of Thomas Joyce and his son, James Joyce. Church and clockmakers 1849-1850 (D). Joyce & Son, Watch (and Church turret and spring clock makers and general dealers). 1851 (D). It was during this partnership that the clock was made for St. Alkmund's church, Whitchurch, 1849.

Joyce, James   Whitchurch, 1821-1883
Clock and watchmaker, silversmith and cutler, High Street. 1856-1863 (D). Watch and clockmaker, 40 High Street. 1868-1875 (D). Son of Thomas and Charlotte Joyce. Bapt. at Whitchurch, 1821. Watchmaker and maltster (employing two men), High Street. 1871 (C). Died, unmarried, 1883.

Joyce, John Barnett   Whitchurch, 1826
Clockmaker. Son of Thomas and Charlotte Joyce, and brother to James Joyce. Appear to have worked at Bradford, Yorks. for a number of years. Turret clock manufacturer, employing six men and two boys, St. John Street. 1871 (C). Children by Ellen Roberta, his wife: Walter Conway, c. 1855; Edith, c. 1857; John Barnett, c. 1861; Arthur, c. 1863; and Mabel, c. 1867.

Biographical List of Clock and Watchmakers 89

Jookes, Philip   **Ludlow, 1542**
'1542. Item, to Phelip Jookes for mendynge the cloocke and the weyche. iiijd.' (CW, St. Laurence, Ludlow).

Kaye, Joseph B.   **Market Drayton, 1871**
'Joseph. B. Kaye, lodger, Stafford Street. Unmarr: age 22, Watch & Clock jobber, born Durham' (C, 1871).

Keeling, John   **Roddington, 1692-1740**
Turret-clock maker. Known to have made a turret clock for 'Whitehall', Shrewsbury, in 1692. (*SC,* 17 February 1871.) Repaired the clock of St. Peter's, Cound, in 1696 and again in 1697 (CW). He entered into an agreement with the Churchwardens of Cound: 'May the 3rd 1697. It is allsoe agreed that John Keeling of Roddington shall be pd 4 shill: p Ann by the sd Wardens if he doe & shall from time to time mend & keep the sd Clock in good & sufficient Order & Repair' (CW). He attended the clock at Cound from 1696 until 1725 (CW). He married Martha Davis of Uppington in 1709. In 1733 he was paid two shillings and sixpence for cleaning the clock of All Saints church, Berrington. There appear to be two makers of this name, both of Roddington, for the registers record the deaths of : 'John Keeling, clockmaker buried 6. Dec: 1734' and of 'John Keeling, clockmaker buried 25. March. 1740.'

Kelvey, Rebecca   **Shrewsbury, 1849-1851**
(*See under* Rebecca Warner.)

Kelvey, James   **Shrewsbury, 1849-1850**
Married Rebecca Warner, widow in the May of 1849. A serious attack on his wife resulted in his being committed to the County Gaol, where he committed suicide by hanging. '1850. 25. Feb: James Kelvey, watchmaker of Mardol, age 34, died in Gaol. buried.' (St. Michael's R.)

Kent, John   **Shrewsbury, 1865-1875**
'J. Kent. (Late Nightingale) Silversmith and Jeweller, Respectfully intimates to his numerous patrons that his Stock is now replete with every Novelty, consisting of Costly & Elaborate

Designs in Jewellery, Silver, & Electro-Plate' (*SJ*, 24 May 1865). 'KENT. New Market Place. Established 1841' (Advert.: *EJ*, 1875). 'J. KENT. 38, High Street. Watch Manufacturer and Goldsmith. A Varied and Elegant selection of Articles suitable for PRESENTATION: Clocks, in Ormolu, Bronze, and many other artistic devices. Jewellery & Plate of every description. WATCH, CLOCK AND JEWELLERY REPAIRS SPEEDILY EXECUTED' (*SJ*, 30 October, 1867).

King, James Shrewsbury, 1768
'James King, Shrewsbury. an. 1768. Watch' (Baillie).

King, Joseph Shrewsbury, 1755
'Joseph King, Shrewsbury. an. 1755. Watch' (Baillie).

King, R. Shrewsbury, n.d.
Watch movement: 'R. King. Shrewsbury' Enamel face (CHM).

King, Thomas Shrewsbury, 1763
'Thomas King, Shrewsbury. an. 1763. Watch' (Baillie).

King, William Shrewsbury, 1774
William King, Shrewsbury. an. 1774. Watch' (Baillie).

Knight, Stephen Ludlow, 1569–1579
'1569. Item payd unto Steven Knight ffor mending of the watch whele of the clock. viijd.' (CW, St. Lawrence, Ludlow.) Stephen Knight was buried at Ludlow, 8 September 1579.

Lamb, John Whitchurch, 1803
'John Lamb, High Street, Whitchurch, age 20, Watchmaker. Sworn a member of the Whitchurch Corp of Volunteers. 4. Sept: 1803.'

Langford, William Ludlow, 1761–1771
'William Langford. 2. June 1761, son of Samuel Langford of Ludlow, apothecary, apprenticed to Pointer Baker. £10. 10s. for 7 years. Freeman 22. Oct: 1770.' (CCRA.) '5. June. 1771. Then William Langford (son of Samuel Langford who was a

Freeman of this Town) was admitted a Freemaster of this
Company to the Trade of a Clock and Watchmaker he paying
a ffine of one pound which he accordingly did to Mr.
Steward Hudson to the use of the company. William Langford' (GHR).
Langford, Wm. Ludlow. 1770 (Britten).

Lashmore, Edward Oswestry, 1863-1891
Watchmaker and jeweller, Cross Street. 1863-1875 (D). Church
Street. 1879. (D). Long-case clock, 8 day, painted dial, moon
movement in arch. (Established 1829.)

Lasseter, Henry Minsterley, 1856
Watch and clockmaker. 1856 (D).

Last, William Bradbury Shrewsbury, 1871-1875
Watch and clockmaker, Pride Hill. 1871-1875 (D).

Launton, Thomas Shrewsbury, n.d.
'Thomas Launton, Salop.—a very old one, with one hand only,
and brass dial' (*Bye-Gones*, 1906, p. 276).

Lawley, Joseph Wellington, 1822/3-1854
Watchmaker, Dun Cow Lane. 1822/3 (D). Dun Cow Lane
and Crown Street. 1828 (D). Crown Street. 1828-1836 (D).
Tan-Bank. 1842-1850 (D). Swine Market. 1851 (D). Watch
movement: 'Jos^h. Lawley. Wellington. No. 5451.' No face
(CHM). Married: 'On Monday, the 28th inst. at St. Chad's,
Mr. Joseph Lawley, watchmaker, to Margaret, Maria, Crump, of
this town' (*SJ*, 30 November 1831). Ten years later, the Census
gives his wife's name as Elizabeth. 'Joseph Lawley, age 60,
Watchmaker. Elizabeth Lawley, aged 65. John Lawley, age 30,
Watchmaker. Vincent Lawley, age 7' (C, 1841). 'Joseph Lawley,
widower, age 71, Watchmaster (Master), of Tan Bank. born
Wellington. John Lawley, son, Unmarried, age 40, Watchmaker,
(Journeyman). born Wellington' (C, 1851). '26.11.1854. Joseph
Lawley, watchmaker from Wellington—Abode—Canal Buildings
—Age 77. buried.' (St. Michael's, Shrewsbury, R.)

Lawrence, Richard Wellington, 1851
Watchmaker, tailor, draper, and pawnbroker. New Street. 1851 (D).

Leach, Edwin Ludlow, 1871
Watchmaker, Corve Street. 'Edwin Leach, marr: age 34, Watchmaker, of Corve Street. born Saddleworth, Yorks: Harriet Leach, wife, age 26, born Liverpool. John Edward Leach, son, age 5, born Dudley. Alfred Leach, son, age 4, born Farnworth, Warwicks: William Leach, son, age 2, born Ludlow. Mary Ellin Leach, dau: age 3, born Ludlow' (C, 1871).

Leigh, Richard Shrewsbury, 1768
—. Leigh, Shrewsbury. 1768. Watch. (Baillie.) Long-case clock, brass face, inscribed: 'Ric: Leigh. Salop'.

Levi, Abraham Wellington, 1840–1846
Watchmaker and Jeweller, New Street. 1840–1846 (D).

Liseter (Lisellen), William Ironbridge, 1828–1836
Clockmaker. 1828–1836 (D). Long-case clock, 8 day, 'W. Lisellen, Coalbrookdale.' Arched, painted dial, cross banded oak case, brass eagle and finials.

Lloyd, Richard Bridgnorth, 1789–1809
Clock and watchmaker. 'Richard Lloyd, watchmaker.' 1789 (UBD). Long-case clock, 30 hour, painted dial, inscribed: 'Ric. Lloyd. B'North.' Oak case.
'CLOCK MAKERS. Wanted. Two or three Journeymen in the above line. Good workmen will receive good Wages and constant Employ, by Applying to Mr. Lloyd, Bridgnorth, and all travelling Expenses paid.' (Advert., *SJ*, 20 December 1809.)

Long, John Hodnet, 1825–1832
Watchmaker. 'On the 10th inst. Mrs. Long, wife of Mr. John Long, watchmaker, Hodnet, in this county' (*SJ*, 23 February 1825). 'On the 10th inst. at Hodnet, aged 43, Mr. John Long, clock & watchmaker' (*SJ*, 23 May 1832).

Biographical List of Clock and Watchmakers

Loseley, Edward Shifnal, 1790–1823
Clock and watchmaker. Edw. Loseley, Shifnal, *c.* 1790 (Baillie). Edw. Losely, clock and watchmaker, Market-place, Shifnal. 1822/3 (D). Long-case clock, Losely, Edward, Shiffnal. About 1790 (Britten). Married a Miss Rathbone of Newport (*SM*, March 1816). 'April. 1821. Ed. Loseby of Shiffnal watchmaker convicted of selling fireworks at his shop, against the Statute of Will: III. Fined £5.' (SQS.)

Loton, John Prees, 1767–1770
'1776. pd John Loton for Cleaning the Clock 0. 3. 0.
1767. pd John Loton for Cleaning and attending the Clock 0. 5. 0.
1770. pd John Loton for Cleaning ye Clock 0. 5. 0.' (CW, Prees).

Lyons, Aaron Oswestry, 1856–1858
Watchmaker, 6 Leg Street. 1856 (D). A Jewish dealer in clocks, watches and cheap jewellery. A mysterious fire broke out upon his premises one night, and at the Salop Assizes of July 1858 he was convicted on the charge of 'setting fire to his dwelling house with intent to defraud an Insurance Company', and sentenced to three years. He was succeeded by T. Clay, watchmaker.

Macklin, Peter Shrewsbury, 1800
'On Tuesday the 2nd inst. aged 75, Mr. Peter Macklin, near 50 years a working clockmaker in this town' (*SJ*, 17 September 1800).

Makin, —. Ellesmere, 1828
Watchmaker, Watergate Street. 1828 (D).

Marsh, G. T. Ellesmere, 1869
Watchmaker. '1869. Jan: 30. aged 26, Frances, Wife of Mr. G. T. March, watchmaker' (*SJ*, 1869).

Marston, William Bishopscastle, 1781
'William Marston, Bishopscastle. 1781. Insolvent. Clock & Watchmaker' (Baillie). (1) long-case clock in heavily carved

oak case, ormolu and silvered dial, 8 day movement, inscribed: 'W. Marston. Bishops Castle.' (2) Long-case clock with square brass dial, inscribed: 'W. Marston. B. Castle.'

Marston, William   Shrewsbury, 1761–1791
Clockmaker. 'William Marston. Shrewsbury. 1761–1791. Clockmaker' (Baillie). High Street 1786 (D). Marston. Clockmaker 1789 (UBD). Long-case clock, oak case, painted dial. Father Time in medallion at top of dial, with the following verse:

> 'To Prove your time
> Watch your thoughts
> Watch your words
> Watch your ways
> Watch your actions
> Watch and Pray.'

Another long-case clock by this maker is of special interest in view of the fact that it has an oval dial. A brass plate on the door is dated 1761, and has the figure of a dragon depicted above it. The oval dial which makes the clock so very unusual is painted white, and like the other clock by this maker, mentioned above, has a medallion at the top of the dial, representing the figure of Hope with an anchor. (The medallion is reminiscent of the style of Pergolesi.) The other decoration is quite interesting as it shows an attempt by the maker of this fine clock to simulate in paint the marquetry work of the Sheraton school, the design being similar in appearance to that found on tea-caddies. There are two smaller dials, one depicting the seconds, and the other one showing the days of the month. The hood of the clock also shows original decoration. The upper spandrels have a blue and gold floral design, covered with glass, while the two lower spandrels are delicately carved. The frame around the oval dial is of beaded work cut in broad and effective style. Although Shrewsbury-made, this clock has an especial Welsh interest to it. It was once in the possession of Daniel Owen, the famous Welsh novelist. The clock figures in his novel, *Rhys Lewis*. The grandmother of the youthful hero of the story journeys to the local fair, and in her absence the young boy takes the works of this clock to

pieces. Then he tries to put it together again. As the hours pass he finds it easier to take to pieces than to put it together again. On the return of his grandmother the clock reveals the boy's secret, for he had forgotten to put back the pendulum, and the clock races like a mad thing; while the culprit uneasily fingers a missing wheel in his pocket.
'Shrewsbury. May. 12. 1791. W. Marston, Clock and Watchmaker, Wyle-Cop, Takes this Opportunity gratefully to acknowledge and return Thanks to the Inhabitants of Shrewsbury and its Vicinity, for the many Favours confer'd upon him since his Commencement in Business, and solicits a Continuance, on Presumption of having given Satisfaction. Every article in the Clock and Watch line got up in the best Manner, Clocks and Watches of all Sorts substantially Repaired; and every Article in the Gold, Silver, Jewellers, and Plated Line, in the neatest Manner and on the most reasonable Terms. Arms, Crests, Cyphers, &c neatly engraved, either Concave or Convex. N.B. Marston scorns to Puff about doing what he is not perfectly Master of himself, a discerning Public will mark the Inference. An Apprentice wanted' (Advert., *SC*, 1791).

Martin, William   Shrewsbury, 1847–1875
Watchmaker, Abbey Foregate. 1847 (PB). The 1851 Census returns him as : 'William Martin, watchmaker, of 25, Abbey Foregate, aged 36, born St. Mary's parish, Shrewsbury. Wife— Rachel, aged 46, Dressmaker, born Holy Cross parish, Shrewsbury'. Watch movement: 'Wm. Martin. Salop. No: 13526.' Enamel face (CHM). William Nicholas Martin, clockmaker, Abbey Foregate. 1875 (D).

Massey, Charles   Ludlow, 1868–1871
Charles Massey, watchmaker, 21 Bell Lane. 1868 (D). 'Charles. K. Massey, age 34, Watchmaker of Bell Lane, born Liverpool' (C, 1871).

Massey, John   Wenlock, 1789–1791
Watchmaker, Wenlock. 1789 (UBD). John Massey, Wenlock. 1791. Watchmaker. (Baillie.)

Massey, John Shrewsbury, 1816-1856
Clockmaker. '1816. 1st. March. (Born 8. Sept:) Edward, son of John & Mary Massey of Tower, Town Walls, clockmaker. Bapt:' (St. Chad's R). '5th. Feb: in her 86th year, Mary, relict of Mr. John Massey, late clockmaker, of this town' (*EJ*, 13 February 1856).

Matthews, Richard Oswestry, 1841-1856
Clock and watchmaker. 34 Bailey Street. 1842-1851 (D). R. Matthews. Oswestry. Early 19th-century watch (Baillie). 'On Monday last, at Oswestry, Mr. Matthews, watchmaker, to Jane, only daughter of Mr. Howell, hairdresser, both of Oswestry' (*SJ*, 21 February 1836). The Census of 1851 returns him as : 'Richard Matthews, marr: age 37, Watchmaker, born Welshpool, of Bailey Street. Jane Matthews, wife age 31, born Oswestry. James. H. Matthews, son, age 12. Jane Matthews, dau: age 8. Emma Matthews, dau: age 10. Mary Matthews, dau: age 5, all born at Oswestry. George Gittens, age 22, watchmaker Journeyman, born Kinnersley.' Richard Matthews is shown as winding and attending the Bailey clock, at a salary of £3 3s. 0d. per annum for the period 26 May 1849 until 9 November 1854. (Corporation Minute Book.) '5th February, after a protracted illness in her 37th year, Jane, wife of Mr. Richard Matthews, watchmaker, Bailey Street, Oswestry; respected by all who knew her' (*SJ*, 1856).

Matthews, James Howell Oswestry, *c.* 1839-1861
Watchmaker and jeweller. Son of Richard Matthew and Jane, his wife. Bapt. Oswestry, *c.* 1839. 'James Howell Matthews, marr: age 22, Watchmaker & Jeweller, born Oswestry. of David Robert's Yard. Emma Matthews, wife, age 25, born Welshampton, Salop. James Henry Matthews, son, age 3, born Birkenhead, Cheshire. Mary E. Matthews, dau: age 2 mths, born Oswestry. Salop. Paul (or Saul) Lloyd, brother-in-law, Unmarr: age 16, (Apprentice watchmaker) born Oswestry' (C, 1861). James Howell Matthews, Bailey Street. Watch and clockmaker. 1856. (D).

Biographical List of Clock and Watchmakers

Matthews, Thomas & Richard   Oswestry, 1836
Thomas and Richard Matthews, Bailey Head. 1836 (D).

Matthews, John   Bishopscastle, 1822-1879
Clock and Watchmaker. Market Cross (Place). 1822/3-1879 (D). 'John Matthews, age 54, marr: Watchmaker, Butter & Market Cross. born Meifod, Montg: Ann Matthews, wife, age 51, born Bishopscastle. John Matthew, son, Unmarr: age 28, Pensioner, Chelsea Hospital. born Bishopscastle. Thomas Matthews, son, age 14, Watchmaker, born Bishopscastle. Richard Matthews, son, age 9, born Bishopscastle. Martha Matthews, dau: age 7. born Bishopscastle' (C, 1851). On the Census of 1841 there were also three more daughters: Emma, aged 12; Sarah, aged 10; and Charlotte, aged 8.

Matthews, Thomas   Bishopscastle, c. 1837-1885
Watchmaker, Church Street. 1885 (D). Apprenticed to his father, John Matthews, in 1851 (C). Bapt. at Bishopscastle, c. 1837.

Matthews, Richard   Bishopscastle, c. 1842-1875
Watchmaker, Church Street. 1868-1875 (D). Son of John Matthews and Ann, his wife. Bapt. at Bishopscastle, c. 1842. Brother to Thomas Matthews above. 'Richard. R. Matthews, age 29, watchmaker of Church Street, born Bishopscastle. Lucria Matthews, wife, age 28, Elizabeth Annie Matthews, dau: age 7. Martha Matthews, dau; age 6, all born at Builth Wells. John Thomas Matthews, son, age 4, born Sandbach, Cheshire. Lucy Matthews, dau: age 2. Charles Henry Matthews, son, age 9 months. Both born Bishopscastle. Edward Sheen Evans, apprentice, born Builth Wells' (C, 1971).

McNiece, John   Bishopscastle, 1861-1875
Watchmaker. 1863-1875 (D). 'John McNiece, age 56, marr: Watchmaker, born Scotland. Sarah McNiece, wife, age 50, Cap-maker, born Gloucestershire' (C, 1861).

McQuinn, John   Market Drayton, 1841
'John McQuinn of Boughey's Yard, aged 30, Barometer Maker. Elizabeth McQuinn, aged 25' (C, 1841).

Meakin, William Ellesmere, 1828-1834
William Meakin, watchmaker, Watergate Street. 1828-1834 (D).

Mew, Samuel Shrewsbury, 1773
'Samuel Mew, Shrewsbury. an. 1773. Watch' (Baillie).

Middleton, John Shrewsbury, 1656
'John Middleton, Watchmaker', one of a large number of Shrewsbury inhabitants named on a writ issued by Oliver Cromwell, to be apprehended and brought before the Justices of the Peace 'for certain contempts and other offences'. Dated 1 May 1656.

Miller, Isaac Newport, 1828-1834
Watchmaker, St. Mary Street. 1828-1834 (D).

Milligan, Thomas Shrewsbury, 1796-1797
Watchmaker of Dog Lane. Admitted a Burgess of Shrewsbury. 1796 (PB). 'A few days ago died, of a decline, Mr. Milligan, junior of this town' (*SC*, 30 June 1797). 'Thomas Milligan, aet 33, buried 18.6.1797' (R).

Millington, Thomas (& Co.) Shrewsbury, 1773-1796
'Millington & Co. Shrewsbury, an. 1773. Watch' (Baillie). Watch, signed 'Millington. Salop.' 1780. (Britten). Thomas Millington of Birmingham, clockmaker, admitted a Burgess of Shrewsbury 1796.

Moody, John Oswestry, 1725-1742
'John moody son of Sarah moody widow hath put himselfe an aprentis to Thomas Nash Clockmaker his time to begin the 8th of June According to ye date of his Indentures 1725.' (CBR). '30.8.1742. John Moody, clockmaker, of Legg St. bur:' (R, St. Oswald, Oswestry).

Moore, John Church Stretton, 1849
Watchmaker. 1849 (D).

Morecock, Daniel   **Shrewsbury, 1775-1781**
Watchmaker. Children by Sarah, his wife: Sarah 1775, bur. 1776; Mary, bur. 1775; and Catherine, bapt. 1778. (R, St. Julian's).

Morgan, Thomas   **Oakengates, 1868**
Watchmaker, of St. George's. 1868. (D).

Morris, Richard and John   **Shrewsbury, 1651**
Blacksmiths who set the clock of St. Michael's, Alberbury in good repair 1651: 'Memorandu that Richard Morris and John Morris, Smithes of the Town of Sheresburie [sic], for and in Consideracon of the Some of Seaventeen shillings unto them in hand paid by Richard Turnor and William Williams, Church-Wardens of ye p'ish of Alberburie, which said Some of 17s. was paid upon the 24th day of October 1651, they the said Richd and John Morris in consideracon of the said Some doe undertake to Sett the Clocke in the Church of Alberburie in good and sufficient order upon the receit of the said Moneis, and the same Clocke thaye by Bargaine and p'mise, and in consideracon of the some of 17s., are to maintaine and keepe from the sd 24th day of October 1651, for and dureinge the Terme and Space of three yeares from thence next ensuinge, and thay or one of them are to come put the said Clocke in order anytime when thay have notice given them that the said Clocke is defective in aney respect.' (R, Vol. II, St. Michael's, Alberbury).

Morris, Robert   **Shrewsbury, 1821**
Robert Morris, Shrewsbury. Early 19th century. Watch (Baillie). Watchmaker, silversmith and cutler, in business nearly 30 years, when in 1821 he handed over his business in the Cornmarket to his nephew, William Baker (q.v.). (SJ, 1 January 1821.)

Morris, John   **Ludlow, 1841**
Watchmaker. 'John Morris, aged 40, watchmaker in the Bull-ring. Mary Morris, aged 30' (C, 1841).

Morris, Thomas   **Albrighton, 1851-1879**
Clock and watchmaker, 1851-1879 (D).

Morris, William   **Newport, 1871**
Clock and watchmaker, of Club Yard, Newport. 'William Morris, of Club Yard, Clock & Watchmaker, marr: age 41, born Weston Jones, Staffs. Ellen Morris, wife, age 26, born Ringley, Lancs. Sarah Morris, dau: age 4, born Newport, Salop. Martha Morris, dau: age 6 mths. born Newport, Salop. James Cheadle, age 16, apprentice Clock & Watchmaker, born Staffs' (C, 1871). Cheadle became a watchmaker on his own account at Shifnal, after his apprenticeship.

Murray, Joseph   **Whitchurch, 1871**
Clockmaker, Beard's Yard. The Census of 1871 returns him as: 'Joseph Murray, aged 37, Clockmaker, born Liverpool. of Beard's Yard. Ann Murray, wife, age 28, born Whitchurch, Salop. Joseph Murray, son, aged 7 mths. born Whitchurch, Salop.'

Nash, Richard   **Shrewsbury ? 1633**
'1633. pd. Rich: Nash for mending the Clocke   13. 4.' (CW, Condover).

Nash, Thomas   **Shrewsbury, 1725-1747**
Clockmaker. Engaged John Moody as an apprentice in 1725, for seven years (q.v.). Also engaged Benjamin Tipton as an apprentice in 1726 (q.v.) (CBR). '13.4.1728. Eleanor, wife of Tho. Nash, bur:.' (R, St. Alkmund's). He obviously married for a second time, as he appears as 'Thomas Nash, clockmaker & Elizabeth his wife,' parties to an Indenture of Lease of property at Longnor, in 1741 (SBL 6460). '13.7.1747. Thomas Nash, aged 56, from St. Chads Buried.' (St. Alkmund's R).

Newall, William   **Cleobury Mortimer, 1789-1812**
Clock and watchmaker. 1789 (UBD). Son: Henry, by Mary, his wife. Bapt. 1794. Served as a Churchwarden at St. Mary's church 1797 (CW). '23.12.1812, William Newhall, aged 50, bur:' (R).

Newall, Henry   Cleobury Mortimer, 1794-1846
Clock and watchmaker. Son of William Newall and Mary, his wife. Bapt. 18 October 1794 at Cleobury Mortimer (R). Clock and watchmaker, 1822/3-1846 (D).

Newnham, Samuel   Shrewsbury, 1832-1834
Watchmaker. 'Mr. Samuel Newnham, watchmaker, appointed Churchwarden at St. Julians. 24th. April. 1832' (*SJ*). 'On the 13th inst. Mr. S. Newnham, watchmaker, Wyle-Cop, in this town, aged 28 years' (*SJ*, 17 September 1834). This business was taken over at this time by James Hanny (q.v.).

Newnes, Samuel   Whitchurch, 1790-1823
Clock and watchmaker of Pepper Alley, 1822/3 (D). Clock and watchmaker. 1789 (UBD). Long-case clock, 8 day, painted dial showing phases of the moon. He engaged John Woolrich as an apprentice in 1794, for a term of five years (SBL 12005). Watch movement, inscribed: 'Sam! Newnes. Whitchurch. No: 189.' No face (CHM). Children by Martha, his wife: Rebecca Higginson 1797; Elizabeth 1800; and Thomas 1803, bur. 1805 (R).

Newton, Isaac   Bridgnorth, 1841
Watchmaker. 'Isaac Newton, age 25, Watchmaker, of Waterloo Terrace. (Not born in Shrps:)' (C, 1841).

Newton, Joseph   Shrewsbury, 1851
'Joseph Newton, age 42, Watchmaker of Castle-Foregate. Born Warwickshire' (C, 1851).

Nicholds, Thomas   Shifnal, 1840-1846
                                             Albrighton, 1851-1875
Watchmaker, etc., of Church Street, Shifnal, 1840-1846 (D). Clock and Watchmaker. Albrighton. 1851-1875 (D).

Nightingale, J. T.   Shrewsbury, 1859-1860
'London Jewellery Establishment. 38, High Street, Shrewsbury. J. T. Nightingale. Proprietor. Nightingale's London & Geneva

Watches, English & Foreign Clocks and Timepieces, Manufactured to order, expressly for his own sale, by the most celebrated makers. etc.etc.' (*SJ*, 20 July 1859).
'J. T. Nightingale, 38, High Street. Watchmaker, jeweller, silversmith, goldsmith, optician.' (*SJ*, 1860). This business was taken over by J. Kent in 1865.

Noke, James   Ludlow, 1718
'James Noke sonne of Wm. Noke was Inrolled an apprentice to Thomas Vernon by the Trade of Clock & Watchmaker & Goldsmith by Ind. bearing date 11. Nov: 1718' (GHR).

Norris, T.   Albrighton, late 18th cent.
'T. Norris, Albrighton. Late 18c. Wall Clock' (Baillie).

Norris, William   Aston (Newport), 1856-1879
Clock and Watchmaker, High Street. 1870-1879 (D). William Norris, watchmaker, Aston (Newport). 1856 (D).

Norrison, William   Newport, 1868
William Norrison, watchmaker, High Street. 1868 (D). Could be a mis-spelling of the surname of the previous man.

Northwood, James   Newport, 1840-1875
Clock and watchmaker. 1840-1875 (D). 'James Northwood, High Street, Clock & Watchmaker, marr: age 59, born Caynton, Salop. Sarah Northwood, wife, age 57, born Brewood, Staffs: Ann Northwood, dau: Unmarr: age 28, born Newport, Salop' (C, 1871). Table-clock, 'James Northwood'.

Norton, Edward   Berrington, 1680
'1680. Itm pd Edward Norton for mending ye Clocke 0. 6. 0.' (CW, All Saints).

Onions, W.   Broseley, 1790
Clockmaker. Long-case clock, face inscribed 'Onions' Broseley.' 1790. Illustrated in *The Grandfather Clock*, by Ernest L. Edwards (pub. 1971).

Ore, Thomas — Tong, Wolverhampton and Birmingham, 1760-1788

Clock and watchmaker. Married Jane Phillips of St. Nicholas, Worcester, at Wellington, Salop, in 1760 (R). 'Ore, Thomas. Wolverhampton and Tong, 1763-79; then Birmingham to 1788. Clock in Birmingham Cathedral; Long-case clocks & watch' (Britten). Children by Jane, his wife (all baptised at Tong): Joseph 1763; Ann 1765, bur. 1766; Anne 1767; Charles 1769; William 1771; Sarah 1773; John 1775, bur. 1776. Sundial at St. Bartholomews, Tong, by Tho. Ore, 1776.

Orme, Michael — Newport, 1764

Watchmaker. Makes a statement (with others) on 12 June 1764 relating to a riot on the previous day, when the gate of the Marsh was broken open by a mob of some 40 people, letting the animals out. (Newport Corporation Records, SRO. 1900/3/20).

Osborne, Joseph — Bridgnorth, 1875-1900; Much Wenlock, 1895-1900

Watchmaker, jeweller and optician; 24 High Street, Bridgnorth, a business which he took over from Robert Hopwood, and which had been established in 1811. Also at Hospital Street, Much Wenlock.

Oswell, —. — Shrewsbury, 1780

'—. Oswell, Shrewsbury. 1780. Watch' (Baillie). Possibly William Oswell, son of Thomas Oswell of Shrewsbury, goldsmith. A William Oswell was admitted a Burgess of Shrewsbury 1788.

Owen, William — Oswestry, 1841-1867

Clock and watchmaker, Cross Street. 1842-1863 (D). 'William Owen. Oswestry. Early 19c. Watch' (Baillie). 'William Owen, of The Cross, marr: aged 50, watchmaker & jeweller, born Llanwchaiarn, Montg: Martha Owen, wife, age 50, born Leominster. William Owen, son, Unmarr: age 21, watchmaker. Samuel Owen, son, age 11. Eliza Owen, dau: age 19. Thomas Owen, son, age 16, watchmaker. Margaret Owen, dau: age 14. Fanny. M. Owen, dau: age 10, all born at Oswestry' (C, 1861). Shown in Corporation minutes as winding and attending the Bailey and Cross Market clocks, at a salary of £5 per year, from 9 November 1854, still receiving salary on the 10 November 1862.

Owen, William and Thomas   Oswestry, 1868–1875
Clock and watchmakers. Sons of William Owen above. Took over father's shop at The Cross. 1868–1875 (D). Later the business appears as William Owen. 1879 (D). Watchpaper: 'W. Owen, Clock & Watchmaker, Jeweller & Silversmith. Cross Street. Owestry. Wedding Rings' (CHM). Watchpaper, as above, but including: 'Optician & Cutler', with a delightful picture of the shop (CHM).

Palmer, Elias, Bailey   Madeley, 1868–1871
Watchmaker and working jeweller. 1868–1871 (D).

Palmer, Joseph   Ludlow, 1851–1856
Watchmaker, High Street. 1856 (D). 'Joseph Palmer, marr: aged 20, Old Street. Jeweller & Watchmaker, born Kidderminster. Elizabeth Palmer, wife, age 23, born Haughton ? Radnorshire' (C, 1851).

Palmer, Thomas   Ludlow, 1871–1900
Clock and watchmaker, 9 Bull Ring, established upwards of 15 years. 1895 (D). Bull Ring. 1891–1900 (D). 'Thomas Palmer, age 24, marr: Watchmaker, of Broad Street. born Coventry. Hannah Palmer, wife, age 25. born Coventry. Eunice Palmer, dau: age 3. born Coventry. Florence Palmer, dau: age 1. born Ludlow' (C, 1871).

Palmer, William   Ludlow, 1667–1668
'1667–1668. Itm payd to Wm Palmer for the dyall clocke and other worke   2. 6. 8.' (CW, St. Laurence).

Palmer, William   Cleobury Mortimer, 1822/3
Clock and watchmaker. 1822/3 (D).

Biographical List of Clock and Watchmakers

Payne, William   **Oswestry, 1753-1757**
Watchmaker, of Leg Street. Children by Mary, his wife: John, bapt. 30 September 1753, and David, bapt. 31 January 1757 (R).

Payne, George   **Ludlow, 1737-1809**
Clockmaker. Apprenticed 1737 to Thomas Vernon of Ludlow. '1. October. 1737. The Company met being Quarter Day in the usuall place in the Church and made George Payne who Served a Regular Apprenticeshipp to Mr. Thomas Vernon a Freemaster of This Company to a Clock and Watch-maker He payeing a Fine of Twenty Shillings to the Use of the Company to Mr. Steward Prodgers. Geo. Payne' (GHR). In 1749, George Payne became Steward of the Guild of Hammermen, at Ludlow. '1746. a bill of George Paine for the "watch part of the clock over the new Cross* amounting to £11. 9. 0." was ordered to be paid' (Corp. Accts.) *N.B.—Buttercross. 'George Payne. Ludlow 1743-95. Watch' (Baillie). He married Jane Sharpe on 2 March 1767. She died six years later, and was buried 27 November 1773. He married secondly, some three months later, 1 March 1774, her sister, Elizabeth Sharpe. This marriage lasted longer; Elizabeth Payne was buried 18 October 1801. Eight years later George Payne died: '22. 9. 1809. George Payne, buried' (R).

Payne, George   **Ludlow, c. 1805-1868**
Clock and watchmaker. The Narrows. 1842-1846 (D). Bull-Ring. 1849-1856 (D). 24 Old Street. 1863-1868 (D). 'George Payne, age 35, Watchmaker, The Narrows. Ellen Payne, age 25, Dressmaker. Elizabeth Payne, age 15' (C, 1841). 'George Payne, aged 46, Conduit Street, Watchmaker, born Ludlow' (C, 1851). 'On the 11th inst. at St. Lawrence's, Ludlow. Mr. George Payne, watchmaker to Mrs. Bowen, widow of Mr. T. Bowen, both of that town' (*SN* & *CR*, 17 September 1842). 'George Payne, widower, aged 56, Old street, watchmaker, born Ludlow' (C, 1861). 'Deaths:—Payne. Feb. 29, aged 68, Mr. George Payne, watchmaker, of Old Street, Ludlow' (*SJ*, 1868).

Payne, William   **Ludlow, 1820-1836**
'26. 12. 1820. William Payne admitted as a watchmaker to the

Guild of Hammermen, Ludlow' (GHR). Clock and watchmaker, The Narrows. 1822/3-1828 (D). King St. 1828-1836 (D).

Pearson, James, Molesworth Bridgnorth, 1836-1856
Watchmaker, gilder and dentist, St. Mary Street. 1836-1856 (D). The 1841 Census returns him as: 'J. M. Pearson, aged 45, Watchmaker, St. Mary St. Sophia Pearson, age 35, Dressmaker. Theophilus Pearson, age 15, Violinist. None born in this county.'

Pedley, Samuel Shrewsbury, 1802
An advert. in the *Salopian Journal,* dated April 1802, paints a sad picture of this man: 'Whereas Samuel Pedley, ClockMaker is under an agreement to Robert Webster of this Town for near Two years to come, notwithstanding which he scarcely works one-third of his Time and that without any just cause but the effect of Idleness and Drunkeness. This is therefore to warn all Persons whomsever against harbouring or employing him after this notice, as they will be dealt with as the law directs.'

Pedroni, J. B. Shrewsbury, 1840
J. B. Pedroni, Barometer Maker, Bridge Street. 1840 (D).

Peplow, William Wellington, 1822/3-1855
Shifnal, 1856-1895
'William Peplow, son of William & Hannah, baptised 19th July 1794' (R, St. Clement Dane, London). Apprenticed to the clockmaking trade 1804. Served later in the 90th of Foot, in Ireland, where he married his wife, Catherine Young, at Armagh. Returned to Shropshire, his father's native county, by 1820, as a clockmaker. Clock and watchmaker, Watling St., Wellington. 1822/3 (D). New Street, Wellington. 1828-1856 (D). 'William Peplow, age 56, Watchmaker, of New St, Wellington. born London. Catherine, wife, age 61, born Ireland. Francis, son, age 20, Watchmaker. Ruth, dau: age 18. Samuel, son, age 16. All children born Salop' (C, 1851). William Peplow, Market-Place, Shifnal. 1856-1891 (D). Aston St., Shifnal 1895 (D). 'William Peplow, age 76, Clock & Watchmaker, of

Market-Square, Shifnal. born Blackfriars, London. Catherine, wife, age 81, born Antrim, Ireland' (C, 1871). Seventy-two years a Wesleyan Methodist. Died at Shifnal, 14 March 1895, aged 100 years and 8 months. All three of his sons, Francis Young Peplow, Samuel Kirk Peplow, and William Peplow followed in their father's trade and became watchmakers.

Peplow, Francis Young Ironbridge, 1830–1887
Watchmaker, son of William and Catherine Peplow, born 2 April 1830 at Wellington. Still living with his parents 1851 and 1861 (see Census). 'Francis Peplow, The Wharfage, Ironbridge. Unmarr: age 41, Watchmaker' (C, 1871). Salop Road, 1871-1875 (D). Married Emma Jukes, 31 March 1871 at Ironbridge. Died at Birmingham, 3 June 1887.

Peplow, Samuel Kirk Madeley, 1834–1863
Watch and clockmaker, High Street. 1863 (D). Son of William and Catherine Peplow, born 8 April 1834, at Wellington. Still living with his parents in 1851 (see Census). 'Peplow—Trevor. 16th July. at the parish church, Much Wenlock, by the Rev: W. F. Wayne. Mr. S. K. Peplow, Watchmaker, Ironbridge to Emma, third daughter of Mr. Edward Trevor, provision dealer, Much Wenlock' (SJ, 23 July 1856).

Peplow, William II Worcester, 1855–
Clockmaker. Son of William and Catherine Peplow, born Armagh, Ireland. Trained by his father. Set up in business at his birthplace, Armagh, then at Cork, returning to England about 1855, setting up in business at Worcester, then at Stourbridge. A business still flourishes today, five generations later.

Percival, William Ludlow, 1722
Clockmaker. 'Then W^m. Percival was Admitted a ffreem^n. to ye Trade of a Clockmaker having been an apprentice & working at ye trade twelve years or upwards paying as a fforeigner to ye Steward Mr. Jon Hattam ye sume of Tenn pounds to ye use of ye Company. per me Will, Percival' (GHR).

Phillips, Thomas   **Ludlow, 1789-1843**
Watchmaker, High Street. 1828-1846 (D). 'Thomas Phillips, Ludlow 1791-5. Watch' (Baillie). Watchmaker. Ludlow 1789 (UBD). 'Thomas Phillips, age 70, Watchmaker, High Street. Maria Phillips, age 50. William Phillips, age 20' (C, 1841). His son, William, bapt. *c.* 1821 became a clockmaker. 'On the 24th ult. Mr. T. Phillips, watchmaker, Ludlow' died. (*SN* & *CR*, 4 February 1843).

Phillips, William   **Ludlow, *c.* 1821-1851**
Watchmaker and builder. Castle Street. 1849-1851 (D). Son of Thomas and Maria Phillips, bapt. Ludlow, *c.* 1821. 'William Phillips, unmarr: aged 30, Castle Street. Watchmaker & Builder, born Ludlow. Maria Phillips, widow, aged 66, born Tenbury' (C, 1851).

Phillips, William,   **Ludlow, 1782-1828**
Watchmaker. High Street. 1822/3-1828 (D). 'Phillips. W. Ludlow. 1782' (Britten).

Phillips, Samuel   **Oswestry, 1771-1780**
Clockmaker. 'Samuel Phillips, Oswestry. 1771. Clock' (Baillie). Phillips, Sam., Oswestry. 1780 (Britten).

Pitman, Arthur   **Wellington, 1871-1875**
Watchmaker of New Street. 1871-1875 (D).

Plant, Thomas   **Newport, 1793-1799**
Clockmaker. 'Thomas Plant, Newport (Salop) 1795. Clockmaker' (Baillie). Married Ann Rider 29 November 1793. Their son, Robert, buried 21 July 1795. His wife, Ann, was buried 22 April 1799 (R).

Plimmer, Abraham   **Wellington, 1757-1822/3**
'18.10.1757. Abraham, son of Nathaniel (& Elizabeth) Plimer of Elerdine bapt:' (R, High Ercall). 'Abraham Plimmer, Wellington, Salop. 1792. Clock and Watchmaker' (Baillie). 'Abraham Plimer, Clock & Watchmaker. Market-Place, Wellington.' 1822/3 (D).

Plimmer, Nathaniel Wellington, 1750/1-1783
'22nd. February 1750/1 Nathaniel, son of Nathaniel (& Elizabeth) Plimer of Elerdine bapt:' (R, High Ercall). 'N. Plimmer. Wellington (Salop) an. 1783. Watch' (Baillie).

Powell, John Prees, 1791
'1791. 20. April. pd John Powell for repairing the Clock 0. 6. 0.' (CW, St. Chad's).

Powell, William Ludlow, 1861-1885
Clock and watchmaker, Castle Street. 1863 (D), and jeweller, 14 High Street. 1868-1879 (D). 1 King Street. 1885 (D). 'William Powell, marr: age 39, of the High Street, watchmaker, born Nailsworth, Glouc: Emily Powell, wife, age 39, born Ludlow. Walter Crundell, nephew, Unmarr: age 18, Watchmaker's Apprentice, born Ludlow' (C, 1871).

Pozzi, Peter Oswestry, 1828-1850
Barometer maker, Willow Street (next to *The Grapes*). 1844 (D). Also looking-glass maker and toy dealer, Willow Street. 1849 (D). Died 1850, aged 76. Amusing story of him teaching Italian in one lesson, for One Guinea, in Watkins *Oswestry*, p. 178.

Price, Thomas Clun, 1868-1891
Clock and watchmaker, High Street. 1868-1879 (D), and taxidermist. 1885-1891 (D).

Priest, John Beywedley Gunnere, 1591
'1591. The vijth day of november in ye yeare above written paid by the churchwardens to John Priest of Beywedley gunnere the some of xxvjs. viijd for mendinge of ye clock who undertook upon ye sight of a letter sent to him by us churchwardens to come and repair ye same clock for a whole yeare space upon his owne charges of any fault be by his workmanship.   hijs Testebz viz.   I.P.'
'1591. Paid to John Priest for mendinge of ye clock. xxvjs. viijd' (CW, St. Osward's, Oswestry).

Pritchard, Thomas   **Shrewsbury, 1774**
'Thomas Pritchard, an 1774. Watch' (Baillie).

Pritchard, John   **Wem, 1856-1875**
Clock and watchmaker, High Street. 1856 (D). Noble Street. 1863-1875 (D). 'John Pritchard, Crown Street, marr: age 61, Watch & Clockmaker, born Stanton-on-Hineheath. Martha Pritchard, wife, age 45, born Lee, Salop. Ann Pritchard, dau: age 12, Frederick Pritchard, son, age 11; Ellen Pritchard, dau: age 8, all born Hineheath, and James Pritchard, son, age 5, born Wem' (C, 1861).

Pritchard, George   **Madeley, 1795-1802**
'George Pritchard, Madeley Wood. *ca.* 1795. Watch in Shrewsbury Museum' (Baillie). 'Pritchard, Geo; Madelywood. Watch. 1802' (Britten). N.B.— No watch by this maker in Clive House Museum collection, Shrewsbury 1975.

Pugh, Benjamin   **Shrewsbury, 1796**
Watchkey maker, of Hill's Lane. 1796 (PB). Admitted a Burgess of Shrewsbury 1796, but shown as of Birmingham.

Rathbone, John   **Shrewsbury, 1717-1764**
Clockmaker. Son of Robert Rathbone, bapt. 14 January 1716/17 (R). '31.10.1762. Mary, dau: of John & Margaret Rathbon, clockmaker, bapt:' (R, St. Julian's). '22.7.1764. John, son of John & Margarett Rathbon, clockmaker, bapt:' (R, St. Julian's).

Ray, Samuel   **Wem, 1747-1770**
Long-case clock: 'the dial shows the phases of the moon and days of the month' (*BG,* 16 August 1905, p. 107). '13.1.1747. Samuel Ray & Anne Booth, both of this parish, married' (R). Children by his wife, Anne: Samuel 1748; Stephen 1750, bur. 1767; John 1753; Eleanor 1758. His wife, Anne, died 4.8.1758.' Samuel Ray remarried: '14.6.1764. Samuel Ray & Sarah Norton, by lic:'. Children by his second wife: Thomas 1765; Sarah 1768. He died on 27 June 1770 (R).

Biographical List of Clock and Watchmakers

Reynolds, David   **Broseley, 1842**

'David Reynolds, Clock & Watchmaker, committed 24.1.1842, charged with feloniously stealing at Broseley, on the 11th December last, one iron screw spanner, value five shillings, the property of William Barker.' (Cal. of Criminal Prisoners, 1 March 1842.)

Rider, Job   **Westbury, 1761**
                                                              Belfast, –1833

'21 Aug: 1761. Job son of Thomas Rider & Elizabeth bapt:' (R, Westbury). This fine clock and watchmaker, although born in this county did not practise his trade in Shropshire, but is listed here as a native of the county. He settled at Belfast as a watch and clockmaker, invented and made clocks according to a new theory, and constructed a clock which went for 12 years without winding. His obituary, which appeared in the *Belfast Guardian* in 1833, is very fulsome in its praise of this son of Shropshire. 'On the 4th ult. in Belfast, in the 76th year of his age, Mr. Job Rider (a native of Broomhill, in the parish of Westbury, in the county of Salop), a man endowed by nature with singular powers of mind, and a heart animated by the most benevolent and generous principles. In the mechanic arts—in science—in miscellaneous knowledge and general information, he stood prominant, above all competitors, in this province. As an instance of his inventive powers, and of his ability to combine practice with theory, and convert, as it were, speculation into reality, we beg leave to remind our readers, that he constructed a time-teller with his own hands, which went for twelve years without requiring to be wound up by any individual during that long period. The principle on which Rider's clock was made was perfectly new:—the slightest change in the weight of the atmosphere wound up this extraordinary piece of mechanism: and it was a matter of moral certainty, that such a change would take place before the time which its works were calculated to run, independently of that circumstance, could have expired. The rotary steam-engine now at work in this town (Belfast) was his invention. His improvements in hydraulics, and other sciences, are too numerous to be detailed in our columns. He was, indeed, a man of most extraordinary talents, in whom genius & assiduity, imagination & patience, were happily combined. He was kind, gentle, humane—ready to assist less able & less experienced mechanics with his advice, his labour, & his money.—Throughout his own domestic circle, his conduct diffused peace & happiness; and, as a general member of society, he conciliated and secured the well-merited esteem and friendship of every good & worthy man who was acquainted with the even tenor of his blameless life. In short, he was a true philanthropist, because he was a christian.'

'Job Rider. Belfast. 1791-1808. Watch. Partner with R. L. Gardner 1805-1807, and with William Boyd after 1807' (Baillie). The family of Rider were also clockmakers just over the border at Welshpool. The parish register of St. Mary, Westbury, lists amongst the benefactions to that parish:

'A Plate of Brass for a Sun-dial by Thomas Rider of Broomhill put up at the expence of the Parish in the year 1792.'

Ridgway, Josiah Whitchurch, c. 1839-85
·'Josiah Ridgway, aged 32, marr: Clockmaker, born Malpas. Ann Ridgway, wife, aged 32, born Whitchurch. Thomas Ridgway, son, aged 5, born Whitchurch. Robert Ridgway, son, aged 3, born Whitchurch' (C, 1871). Shown on the 1851 Census Return as an apprentice to Thomas Joyce, clockmaker, High Street, Whitchurch, and living in. Shown on the paysheets of J. B. Joyce's, 1884-1885. Son: Bruce Edward Ridgway, born 10 March 1871, apprenticed to same firm 17 March 1884.

Rivolta & Del Vecchio Wellington, 1863
Watchmakers of New Street. 1863 (D).

Rivolta, Francesco Wellington, 1868-1885
Watchmaker, New Street. 1868 (D). Crown Street. 1871-1875 (D). New Street. 1879-1885 (D).

Roberts, Edward Ludlow, 1846
Watchmaker of Raven Lane. 1846 (D).

Roberts, William Oswestry, 1861-1871
'William Roberts, Castle Street, age 45, Watchmaker, born Derby. Hannah Roberts, wife, age 41, Dressmaker, born Mitton, Oxfordshire. William Clifton, Father, Widower, age 71, General Labourer, born Mitton, Oxfordshire. William McTurk, nephew, age 4, born Birmingham' (C, 1871).

Roberto, William A. Oswestry, 1872-1891
Watchmaker, jeweller, optician, 21 Bailey Street. 1879-1891 (D).

Robinson, Edward & Co. Shrewsbury, 1857-
Watchmakers, goldsmiths, silversmiths, jewellers, 9 and 10 The Square, Shrewsbury. These premises were bought in 1857, by Edward Henry Robinson and James Greenleaf Robinson, from the executors of the late William Baker, deceased, watchmaker and silversmith, who had occupied them since 1821 (q.v.). The premises had been occupied by silversmiths since an. 1729.
'NOTICE The Establishment of the Late Mr. William Baker, Market Square. (Established 100 Years) Will Re-open Immediately after the sale by auction. Messrs. Robinson & Co. Jewellers, Silversmiths, Cutlers and Opticians of 52, 53, & 54, Bishopsgate Street Within, London. Respectfully inform the Nobility, Gentry and Inhabitants of Shrewsbury and its environs, that they have taken the above premises and business, which they purpose opening in connexion with their old-established London House, immediately after the Auction. etc etc.' (*EJ*, 3 June 1857).

'A PRESENT FOR THE NEW YEAR'
"Accept my —, this token of affection. Be faithful to its admonition, improve by its warnings, and it will teach you the benefit of punctuality—the value of time."
'London made Gold And Silver watches are perfect time-keepers, and vary in price from Three to Twenty Guineas. A written warranty is given with each one when sold, and any regulation required during a period of twelve months will be done gratis. Messrs. R. guarentee their safety by post, free from charge, to

the purchaser. Repairs done by experienced London workmen, Market Square, Shrewsbury, and 52, 53, & 54, Bishopsgate Street Within, London. The full value given for old Gold and Silver' (Advert.: *EJ*, 13 January 1858). Watchpaper: 'E. Robinson & Co. Shrewsbury. Goldsmiths & Watchmakers. (With a Calendar for 1894)' (CHM).
'15.5.1870. at his residence, Market Square, aged 60. Edward Henry Robinson, jeweller and silversmith' (*EJ*, 18 May 1870). This firm is still in business (1977).

Robinson, Henry   Shrewsbury, 1866-1891
'Henry Robinson. Jeweller, Watchmaker, and Practical Optician. Having left the employ of Mr. E. H. Robinson, of the Market Square, respectfully invites an inspection of his new and well-assorted stock, at LONDON PRICES. Plated Goods re-plated. All repairs done on the premises' (Advt.: *EJ*, 15 August 1866). 'Henry Robinson, Jeweller, Watchmaker, Silversmith and Optician. 15, High Street, Shrewsbury. Respectfully invites an inspection of a well-assorted stock. Repairs done on the premises. Experienced workmen in Watches and Clocks. Apprentices wanted in the working department.' (*EJ*, 1 August 1866.) And at 60, Broad Street, Ludlow.

Henry Robinson & Wells   Shrewsbury, 1891-
Partnership of Henry Robinson (above), and Henry Wells, 15 High Street. 1891 (D).

Wells, Henry   Shrewsbury, 1895-1900
Successor to Henry Robinson and Wells above. Watchmaker, jeweller, silversmith and optician. Chronometer maker to the Indian Government 1896-1900 (D). Carriage clock by 'Wells. Shrewsbury.' Watch movement—enamel face, 'Henry Wells. Shrewsbury' (CHM).

Rodgers, George   Market Drayton, 1840-1871
'G. Rodgers. Watch & Clockmaker, agent to Manchester Fire & Life Office, Church St. 1840' (D). Watchmaker, St. Mary Street. 1842-1846 (D). High Street. 1849-1870 (D). Also ale and porter merchant. 1856-1871 (D).

'George Rodgers, marr: age 52, Watchmaker, employing 2 men, & Ale & Porter dealer, employing 1 man. High Street. born Whitchurch. Eliza Rodgers, wife, age 51, born Marchwiel, Denbighshire' (C, 1871).

Rogers, John   Bridgnorth, 1789-1803
Watchmaker. 1789 (UBD). John Rogers, watchmaker, Overseer of the Poor. 1803.

Rogers, Joseph   Oswestry, 1676
'Pd Mr. Joseph Rogers for Tendinge ye Clocke 0.10.0.' Bailey clock, 1676 (Corp. Accts.).

Rossi, Joseph   Shrewsbury, 1860-1862
'Joseph Rossi, Looking and Chimney Glasses, Barometer, Thermometer and Clock Manufacturer. 24, Mardol. Shrewsbury. Barometers, Thermometers and Clocks, carefully cleaned and repaired. All sorts of Alarum clocks' (*SJ*, 15 August 1860). Clockmaker, Mardol. 1862 (VL).

Rowley, Henry   Shrewsbury, 1796-1842
Watchmaker, St. John's Hill. 1796-1828 (D). Shoplatch. 1834 (D). Children by Elizabeth, his wife: Mary Ann 1796; Harriot 1798; Henry 1802 (R, St. Chad's). 'On the 24th ult at Liverpool, in his 74th year, Henry Rowley, watchmaker, late of Shrewsbury' (*SN & CR*, 30 October 1842).

Rowley, Henry   Shrewsbury, 1802-1822
Watchmaker, son of Henry & Elizabeth Rowley above, bapt. 1802 at St. Chad's. 'On Monday last, aged 22, Mr. Henry Rowley, watchmaker, of this town' (*EJ*, 30 October 1822).

Rowley, William   Shrewsbury, 1829-30
'On the 2nd inst. at Longnor, Mr. William Rowley, watchmaker, to Miss Glover, Dressmaker, both of this town' (*EJ*, 15 April 1829). '1830. June 17. Henry, son of William & Jane Rowley, Cross-Hill, watchmaker. bapt:' (R, St. Chad's).

Russell, James   **Madeley Wood, 1818**
'12.7.1818. John. son of James Russell, watchmaker, baptised' (R).

Salter, Joseph   **Oswestry, 1760–1800**
Clock and watchmaker. Llywds Mansion, Cross Street was occupied by Joseph Salter, who was a member of a very old Oswestry family. He turned his hand to many businesses, one of which was printing, and there is in being *A Collection of Psalms,* printed by him in 1789 (*BG,* Oswestry). Churchwarden of St. Oswalds. 1763. Children by Jane, his wife: Richard 1760; Thomas 1761; Jackson 1763; Sarah 1765. (R). 'Joseph Salter, Oswestry. an. 1783. Watchmaker' (Baillie). 'Lately at Oswestry, in the 75th year of his age, Mr. Joseph Salter, much respected by a numerous & respectable circle of friends & relations' (*SJ,* 5 March 1800).

Salter, Robert   **Oswestry, 1789–1810**
Watchmaker, of Cross Street. Author of *The Modern Angler,* published in 1800, although a watchmaker by trade, purchased a site in Bailey Street, erected a shop (now No. 34), and opened it in the seed business. (*BG,* Oswestry). He was a member in 1800 of the newly-formed 'Oswestry Association for the Prosecution of Felons'. Robert Salter, watchmaker. 1789 (UBD). Children by Gertrude, his wife: Frances 1789; Mary 1791. His wife, Catherine (*sic*) was buried 5 February 1795. Robert Salter, Oswestry. 1795. Watchmaker (Baillie). 'Saturday last, Mr. Salter, watchmaker, Oswestry, to Miss. Deakin, of Trewylan, married' (*SJ,* 23 May 1810).

Savage, Richard   **Shrewsbury, 1698–1728**
Clockmaker. 'Savage De Salop, fecit '98. signature on 30 hour, square dial, Long-case clock' (Britten). Two sons were bound apprentice to him—William in 1700, and Thomas in 1703. A further apprentice—Joshua Johnson was bound in 1709. 'Joshua Johnson hath put himselfe apprentis to Richard Savage. Clockmaker & his time to begine the 16 of November 1709.' A son, Richard, was buried in 1705, while his son, William,

died at the end of his apprenticeship, 5 January 1706/7 (R). Elizabeth, his first wife, was buried 7 March 1722/3 (R). He remarried on 19 October 1726, a Margaret Jones. Two years later he was dead, being buried on 27 June 1728.

Savage, William   **Shrewsbury, 1700–1707**
'William savage puts himselfe An apprentice unto his father Richard Savage Clockemaker for seaven years, his time to Begine the 23rd of June 1700' (CBR). '5.1.1706/7. William, son of Richard Savage, clockmaker, buried' (R).

Savage, Thomas   **Shrewsbury, 1701–1706**
'1701. Dec. 26. Thos son of Richd Savage of Shrewsbury, Clockmaker. to Richd Smyth (crossed out) retorned to Rich. Chandler' (CBR). 'Thomas Savage Son of Rich. Savage Clockmaker putt his Selfe an A prentis to His father for Seven years is time Begining 30 of December 1706' (CBR).

Savage, John   **Shrewsbury, 1812–1848**
Clock and watchmaker. Cross Hill. 1812–1829 (D). Mardol. 1830–1841 (D). Mardol and Abbey Foregate. 1837 (D). Admitted a Burgess of Shrewsbury 1812. 'T. Pickstock Respectfully informs his Friends and the Public in general, that he has taken the Shop and Premises lately occupied by Mr. Savage, WatchMaker, Mardol, which he intends to OPEN in a few Days in the Drug Business' (*SJ*, 3 September 1834). On 15 August 1848 Mr. John Savage died (*SJ*, 16 August 1848).

Seaman, Edward   **Dawley, 1868–1879**
Watchmaker. Dawley Green Lane. 1868 (D). Back Street. Shifnal. 1879 (D).

Season, Thomas   **Ludlow, 1549–1568**
'1549. Item, for mendynge the chymez and the cloke to Thomas Season XXd.
1551. Item, to Thomas Season for his yeres wages for the Chymes and the cloke   viijs.
1559. Payd Thomas Season for the Keypinge of the chymes ijs. vjd.

1563. Item, paid Thomas Season, for mendinge the chymes and the barrelle and jake (jack?) of the clockehouse.   viijd.
1568. Item, paid to Thomas Season for tornynge of the clocke on the secounde tenor' (CW, St. Lawrence, Ludlow).

Shaw, Joseph   Wellington, 1856-1875
Clock and watchmaker. New Street. 1856-1875 (D). 'Joseph Shaw, marr: age 34, Master Watchmaker & Jeweller etc. New Street. born Malpas, Cheshire. Ann Shaw, wife, age 35. born Whitchurch. Thomas Shaw, son, age 12, born Chester. John Walford Shaw, son, age 7, born Wellington. Fanny Rebecca Shaw, daughter, age 2, born Wellington' (C, 1861). 'Death— 1870. Sept. 20. age 16, John Walford, son of Mr. Joseph Shaw, jeweller etc, of New Street. Wellington' (*SJ*).

Sheen, William   Hope, 1863-1875
Clock and watchmaker. 1863-1875 (D).

Sheppard, George   Cleobury Mortimer, 1870-1875
Clock and watchmaker. 1870-1875 (D).

Silvester, John   Ludlow, 1638-1639
'1638-9. Item paid John Silvester for makeing the chymes. 6. 10. 0'.(CW, St. Lawrence, Ludlow).

Simpson, Charles   Newport, 1841
'Charles Simpson, age 45. Clockmaker, of Middle-Row, High Street. Hannah Simpson, wife, age 45. James Simpson, son, age 25, Army. William Simpson, son, age 20, Nailor. Henry Simpson, son, age 15. Samuel Simpson, son, age 12. Thomas Simpson, son, age 10. Emma Simpson, daughter, age 6; Mary Simpson, daughter, age 4. Elizabeth Simpson, daughter, age 1' (C, 1841).

Simpson, Christopher   Shrewsbury, 1827
Watchmaker of St. Alkmunds Place. Children by Henrietta, his wife: twins, Anne and Mary, born 13.October 1827, baptised at St. Alkmund's the same day.

Biographical List of Clock and Watchmakers 119

Smith, John Shrewsbury, 1824-1831
Clockmaker, of Mardol. Children by Mary, His wife: Emily Prudence 1824, and Emily Prudence 1831 (R, St. Chad's). N.B.—Obviously the first child had died previous to 1831.

Smith, John Oldbury, 1842-1846
Watchmaker of Halesowen Street. 1842-1846 (D).

Smith, William Shrewsbury, 1826
Clockmaker of Wyle-Cop. Son, William, by his wife, Susan, 1826 (R, St. Julian's).

Smith, William Ap Robert Oswestry, 1601
'1601. payd to wm ap Robert Smith for making a new spring for the Cloke vjd.' (CW, St. Oswald's).

Smythe, Robert Oswestry, 1591
'1591. Paid to Robt. Smythe for mending of ye Klock   ijd.' (CW, St. Oswald's). N.B.—Could possibly be the father of the previous entry.

Speight, James Tong, 1750-1785
'James Speight, Tong. 1750-1785' (Baillie). 'Speight, James. Tong. 1785' (Britten).

Spenser, John Berrington, 1719
'1719. paid Jn. Spenser for looking to ye Clock. 0. 10. 0.' (CW, Berrington.)

Stafford, —. Ironbridge, n.d.
Long-case clock, 30 hour, dial painted with pink roses and birds, 'Stafford. Ironbridge', mahogany inlaid case, brass-mounted pediments.

Stanton, Thomas Oswestry, 1805-1861
Son of Robert Stanton, cutler of Raven Street, Shrewsbury, bapt. St. Mary's. 16 June 1805. Watchmaker of Cross Street, Oswestry 1836 (D). Died Bailey Street, Oswestry, age 56, 19 September 1861.

Stead, Edward   **Ludlow, 1715-1722**
'25. Junij 1717. Memorand that then Edward Stead was Enrolled haveing Set himselfe an Apprentice to Thomas Vernon to ye trade of a Clock & watchmaker for ye terme of Seven years by Ind dated 7th day of octo. 1715. Edward Stead' (GHR). '26. Decem. 1722. Then Edward Stead was Admitted a ffreem^n to ye trade of a Clock & Watchmaker paying 20s to Mr. Steward Hattam for his ffreedome. Ed. Stead' (GHR).

Stephens, Richard   **Bridgnorth, 1733-1780**
Clock and watchmaker. Made an agreement with the Churchwardens of St. Mary, Alverley, to look after their church clock, on an annual basis. 1759. (CW, Alverley). Children by Elizabeth, his wife: Elizabeth 1733; Ann 1738; Mary 1740; Richard 1743; Susannah, buried 1746; Dicksey 1747; William, buried 1754. Richard Stevens, Bridgnorth. an. 1751-5. Watch. (Baillie.) Richard Stephens was buried 23 August 1780 (R).

Sterry, Robert   **Dawley, 1840**
'Robert Sterry, watchmaker, Dawley Green. 1840 (D).

Stockwell, Thomas   **Cleobury Mortimer, 1849-1885**
Clock and watchmaker, Market Street. 1849-1885 (D).

Stockwell, Thomas   **Church Stretton, 1850**
Clock and watchmaker. 1850 (D). N.B.— Possibly a second shop of the previous man.

Stokes, John   **Bridgnorth, 1765**
Watchmaker. 'John Stokes, Bridgnorth, *ca.* 1765' (Baillie).

Stone, Samuel   **Shrewsbury, 1784**
Samuel Stone, Shrewsbury. an. 1784. Watch' (Baillie).

Stone & Allen   **Shrewsbury, 1823-1834**
Silversmiths and jewellers took over the business on Pride Hill, formerly Hilditch, on 24 September 1823. (Advt.: *SJ*, 22 September 1823). Watchmakers, Pride Hill. 1828-1834 (D).

Biographical List of Clock and Watchmakers 121

Street & Pyke   Bridgnorth, 1778
'Street & Pyke, probably Bridgnorth. an. 1778. Watch' (Baillie).

Street, Richard   Bridgnorth, 1768-1789
'Richard Street. Bridgnorth. 1n. 1768. Watch' (Baillie). Richard Strut (Street?) clockmaker. 1789 (*UBD*). Long-case clock, arched brass dial, moon and stars in arch, eyes of moon move with pendulum, 8 day, 'Richard Street. Bridgnorth'. Oak case with herring-bone oak inlay, hood with swan-neck, brass rosettes and centre brass orb.

Street, Richard   Bridgnorth, 1822-1836
Watchmaker, High Town. 1822/3-1828 (D). St. Mary's Street. 1828-1836 (D).

Studley, Thomas   Ellesmere, 1721
Thomas Studley of Ellesmere, clockmaker, admitted a Burgess of Shrewsbury, 1721.

Super, William   Shrewsbury, 1761
'William Super, Shrewsbury. an. 1761. Watch' (Baillie).

Symon   Wellington, 1619
'1619. Itm payd to Symon of Wellington for Settinge the Chimes in order, & for mendinge ye Clocke and for trussinge of iiij belles. xviijs.' (CW, St. Alkmund's, Whitchurch).

Taylor, Edward   Albrighton, 1855
'21.12.1855. at an advanced age, Frances, relict of Mr. Edward Taylor, clock & watchmaker, formerly of Albrighton in this county' (*SJ*, 1855).

Taylor, William   Bridgnorth, 1743-1781
'William Taylor, Bridgnorth. 1743. d. 1781. Clock & Watchmaker' (Baillie). Long-case clock, 8 day, brass dial, with silvered chapter ring. Mahogany case. Children by Mary, his wife: Joseph 1743; Ann, buried 1743; William 1745; Mary 1747, buried 1749; Margaret 1749; Nathaniel 1751; Christabella 1752; Richard 1754 (R). Long-case clock, 8 day, square brass dial,

oak case. '1755. 27. April. Richard Taylor, buried' (R, St. Leonard's, Bridgnorth).

Tailor, John   Bridgnorth, 1841–1851
Clockmaker 'aged 35, Unmarried. of Whitburn St.' (C, 1841): 'aged 41, Unmarried, Whitburn Street, born Bridgnorth' (C, 1851).

Taylor, Joseph   Albrighton and Wolverhampton, 1836–1846
Joseph Taylor apprenticed to Samuel Underhill, High Street, Wolverhampton, but later in business at Albrighton, subsequently in Cock Street (now Victoria Street), Wolverhampton, next to the *Star and Garter* hotel. In 1836 he moved to Lichfield Street on a site now occupied by the Art Gallery. He died in 1846, but the family business was carried on by his son, William, from a shop in Dudley Street. A very fine 'Taylor' clock has been seen, the lunette depicting a painting of Queen Caroline.

Taylor, Thomas   Ellesmere, 1822–1835
Clock and watchmaker. Scotland Street. 1822/3 (D). Cross Street. 1828–1835 (D). Long-case clock, 8 day, painted dial, oak case.
'ELLESMERE, TO CLOCK & WATCHMAKERS. TO BE SOLD OR LET. For a term of years and entered upon immediately. A Dwelling House and Shop, in a good situation in the town of Ellesmere, Salop, also to be disposed of, the whole of the STOCK-IN-TRADE of a well-established Business of a CLOCK and WATCHMAKER lately deceased. For further Particulars apply (if by Letter, Post-paid), to Mrs. Taylor, on the Premises; or to George Salter, Solicitors. Ellesmere'. (*SJ*, 9 December 1835).

Thomas Richard   Shrewsbury, 1824–1835
'30.1.1825. Mary, Anne daughter of Richard & Anne Thomas of St. Alkmund's Square, watchmaker' (R). 'Jane, dau: of Richd & Ann Thomas of Barker St, clockmaker bapt: 1835' (R).

Biographical List of Clock and Watchmakers

Thomas, Richard   Church Stretton, 1822/3
    Watchmaker. 1822/3 (D).

Thomas, Richard   Oswestry, 1836
    Watchmaker of Willow Street. 1836 (D).

Thompson, Thomas   Shrewsbury, 1768
    'Thomas Thompson. Shrewsbury. an. 1768. Watch' (Baillie).

Thompson & Williams   Ludlow, 1870-1875
    Watch and clockmakers of Tower Street. 1870-1875 (D).

Thornton, Thomas. H.   Much Wenlock, 1868-1875
Watchmaker & earthenware dealer, Hospital Street. 1868 (D).
    Watchmaker, Hospital Street. 1871-1875 (D).

Tipton, Benjamin   Ludlow, 1726-1796
'1.8.1726. Benjamin Tipten hath putt himself an apprentres to Tho: Nash Clockmaker for seven years is Time begins ye first day of August according to the date of is Indentuers 2s. 6d.' (GHR). 1747, he became Steward of the Guild of Hammermen, Ludlow. '13.9.1743. Benjamin son of Benjamin Tipton & Anne bapt' (R). '15.11.1796. Mr. Benjamin Tipton
    buried' (R).

Tomley, John   Oswestry, 1680
Gunsmith, of Cross Street, his son, Robert Tomley was also a gunsmith in Church Street.'1680. Pd John Tomley, Smith, towards setting up ye Clocke   01. 0. 00.' N.B.—The Bailey
    Clock. (Oswestry Corporation Accts.)

Tranter (Traunter)' Thomas   Shrewsbury, 1729-1784
    Clock and watchmaker, of the High Street.
'May ye 5. 1729. Tho. Traunter son of Robert Traunter of Welshpoole In Moungomry hath putt hi self an a prentice to George Birchall Clockmaker for seven years. is time begins according to deate of Is Indenture. Rec^d 00. 2. 6.' (CBR). Long-case clock, 8 day, brass dial, inscribed 'Traunter. Salop.' mahogany banded oak case. Children by Elizabeth, his wife:

Jane 1750, bur. 1753; Robert 1753, bur. 1753; Thomas, aged 13, bur. 1755; Thomas and Jane (twins) 1755, bur. 1755; Jane 1758, bur. 1758. Thomas Traunter, watchmaker, High Street. 1774.' (BL). '4.8.1770. Elizabeth, wife of Thomas Traunter, Watchmaker, aged 53, buried' (R). '21.12.1784. Mr. Thomas Traunter, aged 68, buried' (R, St. Alkmund's).

Torkington, Jeffrey　　　　　　　　　**Whitchurch, 1789–1799**
'Jeffrey Torkington, Whitchurch (Salop). 1795. C.' (Baillie). 'Jeffr. Torkington, Clockmaker. 1789' (*UBD*). '13.6.1799. Thomas, son of Jefre & Mary Torkington baptised.'

Trentham, Thomas　　　　　　　　　**Shrewsbury, 1774–1779**
Thomas Trentham, watchmaker, Freeman of the Borough of Shrewsbury 1774. 'Thomas Trentham, Shrewsbury. an. 1779. Watch' (Baillie).

Turford, James　　　　　　　　　**Ludlow, 1870–1885**
Watchmaker, Raven Lane. 1870–1875 (D). Castle Street. 1879 (D). Churchyard. 1885 (D).

Turford, James Benjamin　　　　　　　　　**Ludlow, 1891–1900**
Watchmaker and woodcarver, 16 Raven Lane. 1891 (D). Churchyard. 1895(D). 28 Lower Broad Street. 1900 (D). Possibly the same man as the previous entry; or a son.

Turner, Henry C.　　　　　　　　　**Oswestry, 1861**
'Henry C. Turner (Lodger with Mary Surma, widow of Cross Street) Unmarried, age 31. Watchmaker. born Holy Trinity, Colchester' (C, 1861).

Underhill, Thomas　　　　　　　　　**Albrighton, 1781–1830**
5 June 1781. Thomas Underhill, bachelor and Anne Baddeley, spinster, married by license (R). His wife, Anne, was a daughter of John Baddeley, clock and watchmaker of Tong and Albrighton, and Martha his wife. Bapt. 1756. Children by Anne, his wife: William 1782; George 1784; John 1785; Thomas 1790; Elizabeth 1792; Martha 1795; Sarah1797; and Samuel 1801

(R). Thomas Underhill died 28 May 1830. His son, William, became a watchmaker at Newport.

Underhill, William   **Newport, 1782-1836**
Watchmaker. Son of Thomas Underhill, watchmaker of Albrighton, and Anne, his wife. Bapt. 4 April 1782(R). William Underhill, High Street. 1828-1836 (D). Watch movement: 'Underhill. Newport.' Enamel face (CHM). William Underhill, watchmaker. 1832 (VL).

Vernon, Thomas   **Ludlow, 1701-1740**
Possibly the 'Vernon, Tho. 20. May, 1701. apprenticed to Robert Halsted. 7 years. Free 2. August. 1708' (CC). '22nd of March 1711. At a Meeteing then held for the Company ffees of Smyths & others It was agreed by the Six Men and the Rest of the Company then p'sent that Thomas Vernon be Admitted a ffree master to a Clocke & Watchmaker a Member of ye sayd Company he paying to the present Steward Crow three pounds for ye use of the company (fforty shillings being already payd to George Cooke the former Steward towards his fine for his ffreedom this being five pounds. Tho: Vernon' (GHR). '17.1.1710/11; Thomas Vernon & Winifred Noke both of this parish' (R). Richard Felton apprenticed to Thomas Vernon, 1 May 1711. Admitted a freeman, 17 June 1718 (GHR). Edward Stead apprenticed to Thomas Vernon, 7 October 1715. Admitted a Freeman, 26 December 1722 (GHR). James Noke (a kinsman of his wife) apprenticed to Thomas Vernon, 11 November 1718. Admitted a Freeman, 11 April 1726 (GHR). Long-case clock, brass dial, inscribed 'Tho. Vernon. Ludlow', single hand, oak case, surmounted with three ball and spike finials. Sundial in St. Lawrence's churchyard, Church Stretton (brass), inscribed: 'Tho. Vernon. Ludlow' (gnomen missing). '28.11.1738. Winnefred, wife of Mr. Tho. Vernon buried' (R). '9.5.1740. Mr. Thomas Vernon buried' (R).

Vickers, George Henry   **Wellington, 1861**
                                      Oakengates, 1856-1875
Clock and watchmaker, jeweller, New Street, Oakengates,

1856–1875 (D). 'George, Henry, Vickers, marr: aged 27, Watchmaker, Jeweller etc. of New Street, born Wrockwardine. Claudio, Beata, Vickers, wife, aged 22, born Harmer Hill. George, Arkinstall Vickers, son, aged 1, born Wellington' (C, 1861).

Waler, William   Much Wenlock, 1828
Clock and watchmaker, 1828 (D).

Walker, Charles   Ludlow, 1789–1795
Watchmaker. 1789 (UBD). 'Charles Walker. Ludlow. 1791-5. L.c. clock. Virginia. M.W.' (Baillie).

Walker, William   Shrewsbury, 1841–1870
Clock and watchmaker, aged 25. Mardol. 1841. (C.) William Walker. Mardol. 1841 (VL). Wall-clock, inlaid flower design in mother-of-pearl on black background, inscribed: 'Walker. Shrewsbury'. William Walker, watchmaker. Mardol. 1841–1847 (PB), watchmaker, Mardol. 1842–1846 (D). Market Square. 1849–1856 (D). Long-case clock, in oak and mahogany case, painted dial, inscribed 'Walker, Shrewsbury', 8 day, scroll hood. Watchpaper: 'Walker. Clock & WatchMaker. Mardol. Shrewsbury. Gold Rings &c.' dated 1842 (CHM). Watchpaper: 'Walker, Clock & Watchmaker. Market Square, Shrewsbury. Gold Rings etc.' (CHM). 'William Walker, aged 49, watchmaker. Dogpole. born Hull, Yorks. George Walker, son, aged 10, born Shrewsbury' (C, 1861). Children by Catherine, his wife: William Charles 1849; John Oxborough 1852; Catherine Mary 1854; John Oxborough 1855; Alfred Oxborough 1857; Frederick Oxborough 1860; Albert Oxborough, bur. 1865 (R).

Walters, William   Clun, 1842–1856
Watchmaker. 1842–1856 (D).

Walker, William   Shrewsbury, 1827–1829
Watchmaker of Cross Hill. Children by Mary Ann, his wife: Caroline 1827, and George 1828. (St. Chad's R).

Warner, John — Shrewsbury, 1840-1844
Succeeded to John Savage, watchmaker, Mardol. John Warner, watchmaker, Mardol. 1842-1844 (D). '3.2.1840. Ann Warner, wife of John Warner, watchmaker of Castlegates. age 30, buried' (St. Mich. R). He re-married again, for his widow Rebecca took over his business on his death. '6.5.1844. John Warner, watchmaker of Mardol, age 39. buried' (St. Mich. R).

Warner, Rebecca — Shrewsbury, 1844-1851
Watchmaker. Widow of John Warner above. Watchmaker, Mardol. 1849-1851 (D). Married James Kelvey, watchmaker, May 1849, soon after which he made a vicious attack upon her, resulting in him being put into the county gaol, where he committed suicide.

Weatherby, Thomas — Market Drayton, 1861-1900
'Thomas Weatherby, age 19, unmarr: of Church Lane, watchmaker's apprentice, born Drayton in Hales' (C, 1861). 'Thomas Weatherby, marr: age 29, Clock & Watchmaker, Church Street, born Drayton, Salop. Eliza Weatherby, wife, age 20, born Tamworth, Staffs.' (C, 1871). Watch and clockmaker, Church Street. 1879-1900 (D).

Webb, Joseph — Shrewsbury, 1851-1875
'Joseph Webb, aged 17, Clock & Watchmaker of 112, Frankwell, born Ludlow. son of Serg. Major Samson Webb, Militia Staff. (also of the same address, & born Ludlow)' (C, 1851). Clock and watchmaker, 56 Mardol. 1863-1879 (D). Joseph Webb, watchmaker, Mardol. 1862 (VL). Watch movement: 'Jos^h. Webb. Shrewsbury. No. 42906.' Enamel face (CHM). Children by Jane, his wife: Mary Elizabeth, bur. 1859, aged 4; Martha Elizabeth, bur. 1869, aged 4 months; Joseph Henry, bur. 1870, infant; Alfred Joseph 1872; Hannah Clara 1871; Ernest John 1875.

Webb, William — Wellington, 1726-1786
Clock and watchmaker. 'William Webb, Wellington, *ca.* 1750. Watch mount, Buckley Collection.' (Baillie). 'January. 27. 1796. TEN GUINEAS REWARD. Whereas early on Sunday Morning,

the 24 inst. A ROBBERY was committed in the House of
Mr. Wm. EDWARDS, of ARLSTON in the Parish of WELLING-
TON, and County of Salop and the following Articles stolen:
one silver Watch, Maker's Name Wm. Webb, Wellington, numbered 719, and Three Five Guinea Bills, Wellington Bank, and
One Guinea in Gold. The Person suspected is a man servant,
who lived with Mr. Edwards, and absconded early the same
morning; he goes by the name of JOHN DERRICOT, is about
26 years of age, 5 feet 8 or nine Inches high, long Black strait
Hair. Pale Complexcon, an Eruption on both Cheeks, and a
large black lump on the lower part of the Shin of his left leg,
to be seen thro' the stocking: had on when absconding a New
Smock Frock, with a large Blue Coat over it, a Black silk
handkerchief round his neck; has with him a Pistol, and four or
five Shirts, supposed to have Frills at the Breast, but picked off:
is thought to be a Deserter. Whoever apprehends the said
offender, so that he may be convicted, shall receive the above
reward from the Wellington Association for Prosecuting Felons,
over and above what is allowed by Act of Parliament. By
order of the Committee— John Wood, Attorney,
Treasurer, Wellington. Jan. 25, 1796.'
'2.12.1726. William, son of William & Mary Webb, bapt:' (R).
'27.12.1750. William Webb & Mary Dale married' (R). Children
by Mary, his wife: Anne 1751; Thomas 1755; John 1757; and
Eliza 1759 (R). William Webb, Wellington. Watch movement:
'No, 1832,' verge escapement. c. 1765. Without a case, fusee
chain gone (LVM). Long-case clock, brass face. '8 January.
1786. William Webb, aged 58, buried' (R).

Webster, James Shrewsbury, 1740-1799
Clock and watchmaker, Mardol. Established 1740. '20.10.1745.
James Webster, of the parish of St. Chads, & Elizabeth Rathbone, of the parish of Holy Cross, Salop, married by License'
(R, St. John the Baptist, Stapleton). Children by Elizabeth,
his wife: James 1746; Henry 1748, bur. 1766; Elizabeth 1750;
Mary 1753, bur. 1755; Robert 1755; Margaret 1757, bur.
1757; John 1761 (St. Chad's R). In 1774 he supplied a new
clock for Old St. Chad's, Shrewsbury (CW). James Webster,
clockmaker, Mardol. 1768 (BL). Long-case clock, 30 hour,

brass face, oak case, inscribed: 'Jas. Webster. Salop' (CHM). '8.8.1779. Elizabeth, wife of James Webster. buried, aged 60' (St. Chad's R). '11.11.1799. James Webster, aged 85, buried' (St. Chad's R). '13.11.1799. Died.—On Friday last, at the advanced age of 85, Mr. James Webster, formerly a clockmaker, of this town' (SJ). Succeeded by his son, Robert Webster.

Webster, Robert   **Shrewsbury, 1755-1832**
Clockmaker of Mardol. Son of James and Elizabeth Webster. Bapt. 10 January 1755 (St. Chad's R). Admitted a Burgess of Shrewsbury, 17 May 1796. Married Martha Baddeley, dau. of John Baddeley, clockmaker, of Tong and Albrighton (q.v.) in 1776. Children by Martha, his wife: Martha 1777; John Baddely 1779; Robert, c. 1781 (R). Robert Webster established a large manufactory in Mardol in 1783. He made several turret clocks for buildings in Shrewsbury—Millington's hospital, St. Alkmund's church, and the County gaol, and clocks of his manufacture were set up in the churches of Alberbury, Chirbury, Clunbury, Ellesmere, and Worthen. Long-case clock, brass face, inscribed 'Robert Webster. Salop', 30 hour. A handbill dated 5 June 1797 offers two guineas reward for the recovery of 'a silver watch, maker's name "R. Webster. Salop. No. 1007".' He was an inventive genius of no mean ability, and is reputed to have made a clock which required winding only once a year. He also presented to Queen Caroline three elegant spinning wheels, two of which were his own invention. His inventive brain turned to the washing of clothes, and he invented and patented a Washing Machine in 1792, and a mangle in 1812. 'ROBERT WEBSTER Clock and Watch Maker, Shrewsbury. Informs the Public that he has taken the House and Shop formerly occupied by Messrs. Cooke & Son, Grocers, Mardol, and hopes by a strict Attention to every order he may be honoured with, to merit their future Favours' (SJ, 23 November 1814). In 1817 he handed his business over to his son, John Baddeley Webster, but not to resign his business life, but, at the age of 62, to open a new business: 'Wholesale and Retail Brush Manufactury, Mardol'. His wife died in 1832: 'On the 6th inst. aged 79, Martha, relict of the late Mr. Webster, of Mountfields, Frankwell, in this town, & eldest daughter of

Mr. Baddeley, formerly of Albrighton, in this county' (*SJ*, 7 March 1832). His son, Robert, died before him, at an early age: 'Saturday last, in the prime of life, after a lingering illness, which he bore with the greatest patience & resignation, Mr. Robert Webster, jun: ironmonger, of this town' (*SJ*, 19 April 1809). A month before his wife, Robert Webster passed away: 'Deaths.—7th instant at Mountfields, Frankwell, in his 78th year, Mr. Robert Webster, formerly clockmaker and ironmonger of this town' (*SJ*, 10 February 1832).

Webster, John Baddeley Shrewsbury, 1779-1829
Clockmaker. Son of Robert and Martha Webster. Bapt. 10 February 1779, at Shifnal (R). Set up as a clockmaker on his own— in Dog Lane (now Claremont Street) 1796; High Street, 1812-1814, and succeeded to his father's business in Mardol in 1817. Married Mary Roberts at Baschurch by license, 12 November 1804 (R). Admitted a Burgess of Shrewsbury, 1807. A son, James, born 1805. His wife, Mary died in 1837. 'On the 16th inst. in his 51st year, after a severe illness, Mr. John Webster, clockmaker, Mardol, in this town' (*SJ*, 24 December 1829).

Webster, James Shrewsbury, 1805-1871
Clockmaker. Son of John Webster and Mary, his wife. Bapt. 1 January 1806 (R). Admitted a Burgess of Shrewsbury, 1830. He moved the family business away from Mardol where it had been established so long, and is noted in directories 1836 onwards as at Milk Street (sometimes referred to as Belmont). Parish clerk of St. Chad's, 1835 onwards. His son, James Baddeley Webster, an accountant, was also organist of St. Julian's. '1871. August 17. at Belmont, in this town, in his 66th year. James Webster' (*SJ*, 1871). '19.8.1871. James Webster, (Parish Clerk) of Belmont. aged 65. buried' (St. Chad's R). Thus ended a family concern dating from 1740, through four generations.

Webster, John Ironbridge, 1796
'John Webster of Ironbridge, Clockmaker, admitted a Burgess of Shrewsbury 1796.'

Welch, Henry   **Shrewsbury, 1867**
'7.2.1867. William, Thomas, son of Henry & Ann Welch. Barker St. Watchmaker. bapt: (born 19.5.1866)' (St. Chad's R).

Wellings, William   **Ellerdine, 1832–1856**
Clock and watchmaker. 1832 (VL). 1852–1856 (D).

Wheeler, Cornelius   **Bridgnorth, 1770–1842**
Watchmaker and jeweller. Son of Thomas and Sarah Wheeler. Bapt. 20 October 1770 (St. Mary's R). High Town. 1822/3 (D). High Street. 1828–1842 (D). Sons by Anne, his wife: George 1792; Charles 1797. 'Cornelius Wheeler, age 70, watchmaker High St. Ann Wheeler, age 68' (C, 1841). 'On the 26th inst. at Bridgnorth, aged 73, Mr. Cornelius Wheeler, watchmaker, universally respected by his friends & neighbours' (*SN* & *CR*, 29 January 1842).

Wheeler, George   **Much Wenlock, 1792–1850**
Clock and watchmaker. Son of Cornelius and Ann Wheeler of Bridgnorth. Bapt. 30 November 1792 (R). Watchmaker of Willmore Street. 1828–1834 (D). Barrow Street. 1836 (D). Watchmaker and landlord of the *Punchbowl*, 1840 (D). Spittle Street. 1842–1846 (D). High Street. 1849–1850 (D). Two long case clocks seen, 30 hour (but with two winding holes on dial to make them appear 8-day clocks), painted dials, 'George Wheeler. Much Wenlock'. Inlaid cases.

Wheeler, John   **Much Wenlock, 1851–1856**
Clock and watchmaker of the High Street. 1851–1856 (D). A son of George above.

Whiston, Joseph   **Newport, 1828–1863**
Clockmaker and watchmaker, High Street. 1828–1863 (D). 'Joseph Whiston, marr: age 64, Middle Row, High Street, Watchmaker (Master) born at Patishull, Staffs. Fanny Whiston, wife, age 55, born Sambrook' (C, 1861).

Whiston, Thomas   **Newport, 1851–1891**
Clock and watchmaker, Upper Bar. 1851–1856 (D). High Street. 1863–1891 (D). 'Thomas Whiston, lodger, Upper Bar, age 30,

Clockmaker, born Newport' (C, 1851). 'Thomas Whiston, High Street, Clock & Watchmaker, marr: age 50, born Newport, Salop. Elizabeth Whiston, wife, age 39, born Newport. Thomas. J. Whiston, son, age 12; Fanny Whiston, dau: age 8; George Whiston, son, age 5. All born at Newport, Salop.' (C, 1871). Thomas was probably the son of Joseph and Fanny Whiston of the previous entry.

Wicksteed, Charles Oswestry, 1728-1748
Watchmaker and goldsmith, Bailey Street. Children by Alice, his wife: Mary 1728; Ann 1730; Edward 1732; Charles 1739, buried 1742; Alice buried 1748.

Willoughby, Richard Oswestry, 1702
'Aug: 14th. 1702. Pd Richard Willoughby, of Oswestry, for mending the great Clock att Plase is clawdd, which my Master ordered to be done. 1. 0. 0.' (CCA).

Wilson, Henry Wellington, 1851
'Henry Wilson, age 25, New Street, Watchmaker. born London' (C, 1851).

Wilson, William Dawley, 1871-1900
Watchmaker, Chapel Street. 1891-1900 (D). 'William Wilson, Watchmaker, age 37, marr: High Street. born Dawley. Sarah. A. Wilson, wife, age 36, born Little Wenlock. Margaret. A. Wilson, dau: age 12; John. K. Wilson, son, age 10. Sarah. A. Wilson, dau: age 8; Mary. J. Wilson, dau; age 6; Ruth Wilson, dau; age 2; William. H. Wilson, son, age 1' (C, 1871).

Winter, Samson Wellington, 1849-1863
Clock and watchmaker, jeweller, New Street. 1849-1863 (D). 'Sampson Winter, widower, age 62, New Street. born Prussia, Germany' (C, 1861).

Woller, Charles Oldbury, 1840-1846
Charles Woller, watch and clockmaker, Freeth Street. 1840 (D). Halesowen Street. 1842-1846 (D).

Wood, George — Shrewsbury, 1700-1741

'George Wood. ca. 1700. Clockmaker' (Baillie). '1.5.1741. George Wood buried' (R, St. Mary's).

Wood, Richard — Shrewsbury, 1734-1752

Watchmaker, High Street. Son of Isaac Wood of Knutsford, yeoman. Admitted a Burgess of Shrewsbury, 1737. Children by Mary, his wife: Isaac 1735/6; Mary 1737; Anne 1738/9; Richard 1741; William 1743; Elizabeth 1745; Richard, buried 1734. All baptised at the High Street Church, and afterwards entered in the registers of St. Alkmund's (R). 'Richard Wood. Shrewsbury. ante 1745, died 1752. Watch' (Baillie). 'Mathew Holland hath Put him an Aprentefs to Richard Wood for the Space of Seven years according to the Date of his Indenture. Recev'd for the Use of the cumpany 5s.' (CBR, c. 1748). '30.4.1752. Wood, Mr. Richard, Watchmaker, aged 47. buried' (R, St. Alkmund's).

Wood, Mary — Shrewsbury, 1752-

'Mary, widow of Rich. Shrewsbury. From 1752' (Baillie).

Wood, Isaac — Shrewsbury, 1735-1801

Watchmaker. Son of Richard Wood and Mary, his wife, baptised 21 February 1735/6 at the High Street Church. Admitted a Burgess of Shrewsbury 1770. Company Secretary to the 'Salop Fire Office', 1780-1801. The business of the Fire Office was originally carried on at the shop of Isaac Wood, watchmaker, in the High Street, later being transferred to the 'Corn Exchange'. Trustees of the High Street Church 1767. Assignee to Thomas Cook, jeweller, a bankrupt 25/26 July 1794. (SRO, 824/13). 'At Clapham—Mrs. Sarah Wood, widow of the late Mr. Isaac Wood, of Shrewsbury' (*SM,* September 1815). Watchpaper: 'I. Wood. Clock & Watchmaker. High Street. Shrewsbury' (CHM). '19.1.1801. Isaac Wood, aged 65 years buried' (R, St. Alkmund's). A stone in St. Alkmund's churchyard to Isaac, Richard and Mary Wood, reads:

> 'Thy movements, Isaac, kept in play,
> Thy wheels of life felt no decay
> For fifty years at least:
> Till by some sudden secret stroke,
> The balance or the mainspring broke,
> And all the movements ceas'd.'

Wood, James William Shrewsbury, 1875
Watchmaker, of Castlefields. 1875 (D).

Wood, Samuel Sutton, 1698
'Samuel Wood. Sutton (Salop). married 1698. W.' (Baillie.)

Wood, Samuel Ludlow, 1836–1861
Watchmaker. Raven Lane 1836–1849 (D). Bell Lane, 1850–1851 (D). 'Samuel Wood, aged 55, Watchmaker; Mary Wood, aged 45; George Wood, aged 15; Ellen Wood, aged 15; John Wood, aged 13; Emily Wood, aged 7' (C, 1841). 'Samuel Wood, Watchmaker, aged 65, Bell Lane. born Kidderminster, Worcestershire. Mary Wood, wife, aged 58, born Birmingham. George Wood, aged 28, Solicitor's Managing Clerk, born Tenbury. Son. Emily Wood, daughter, aged 16, born Ludlow' (C. 1851). 'Samuel Wood, widower, age 75, Watchmaker, Bell Lane' (C, 1861). Died 1 December 1861.

Wood, Benjamin Ludlow, c. 1831–1861
Watchmaker, 27, Bull Ring. 1868 (D). 'Benjamin Wood, Watchmaker, marr: age 30, born Ludlow. Bull-Ring. Elizabeth Wood, wife, age 24, born Ludlow. Alice Wood, daughter, age 4; Samuel Wood, son, age 3; Herbert Wood, son, age 2 months, all born Ludlow' (C, 1861).

Woodruffe, James Shrewsbury, 1762
'James Woodruffe, Shrewsbury. 1762. Clock & Watchmaker' (Baillie). Long-case clock, 30 hour, brass face, oak-case.

Woodruffe, Thomas Shrewsbury, 1767–1801
Clock and watchmaker, Wyle Cop. 1774 (BL). Children by his wife, Ann: Thomas, Gerrard 1767; Mary 1768; John 1770,

buried, 1816; Richard 1772, bur. 1772; Elizabeth 1773; William and Sarah (twins) 1774; Sarah, bur. 1774; James 1775. (R).
'ThomasWoodruffe, Shropshire. Late 18c. bracket-clock' (Baillie). Thomas Woodruffe, watchmaker. Wyle Cop. 1774 (BL). 5 October. 1801. Thomas Woodruffe buried aet 65' (R, St. Chad's). His wife, Ann, lived on for 54 years after his death: 'Ann Woodruffe (Abbey Foregate) bur: 19. March, 1855. aet 99' (R, Holy Cross).

Woodward, Edmund John   **Bridgnorth, 1856–1860**
Watchmaker, jeweller and silversmith, High Street. 1856 (D). He died, aged 60, 18 April 1860.

Woodward, Maria   **Bridgnorth, 1861–1868**
'26.1.1861. Received information that a tramping watchmaker, James Marsh Hay by name, had absconded from the employment of Mrs. Maria Woodward embezzling 5/- and taking a quantity of tools her property' (*Chief Constables Journal*). Maria Woodward, 24 High Street. 1868 (D).

Woolley, William   **Tong, 1782–1856**
'8.10.1782. William, son of Thomas & Elianor Woolley bapt:' (R). '14.8.1806. William Woolley, bach: & Ann Tagg, spin: both of Tong, marr:' (R). Children by Ann, his wife: George 1807; Mary 1808; Lucy 1810; Ellen 1812. Clockmaker of Tong Hill, Parish Clerk of St. Bartholomew, Tong 1801. Reputed to have been apprenticed to John Baddeley, blacksmith of Tong, who made the Sheriffhales church clock. William Woolley died, aged 74, in the December of 1856.

Woolley, George   **Shifnal, 1807–1836**
'5.7.1807. George, son of William and Ann Woolley bapt:' (R, St. Bartholomew, Tong). Long-case clock, inscribed 'G. Woolley. Shiffnall.' George Woolley, High Street, Shifnal. 1836 (D).

Woolrich, Thomas   **Cheswardine, 1624–1626**
'1624 to Thomas Wolrish for mending ye clocke 2s.
1625 .... ye clock .... 8d.
1626 Disbutsings pd for mending the clocke to Thomas Wollderedg iijs. iiijd.' (CW, St. Swithin, Cheswardine).

Wright, Thomas   **Wellington, 1744**
'Thomas Wright. Wellington. 1744. Clockmaker' (Baillie).

Wynn, William   **Hinstock, 1832–1856**
Clockmaker. 1832 (EL), of Lockley Wood. 1856 (D). Hinstock, 1851 (D).

Young, –.   **Cleobury Mortimer, 1828**
— Young. Clock and watchmaker. 1828 (D).

J. B. JOYCE & CO., LTD., WHITCHURCH

THIS FIRM of high-class craftsmen, whose clocks are to be seen all over the world, have a reputation for the excellence of their turret clocks which remains unsurpassed. Their catalogue, entitled 'High Class Church and Turret Clocks' proudly states their claim to be 'The Oldest Makers of Clocks in the World', a statement which has never been seriously challenged, for the same catalogue points out that they were established in the year 1690.

It is feasible that the George Joyce, apprenticed to Nathaniel Pyne, clockmaker of London, in 1684, was a member of this Shropshire family of Joyce, although the records of the Clockmaker's Company do not give his parentage, nor his place of origin; while it is perhaps significant that a George Joyce is listed on the Shropshire Hearth Tax Roll of 1672.

On 9 February 1691/2, William, son of John and Elizabeth Joyce, of the Lodge (halfway between Cockshutt and Ellesmere) was baptised at Ellesmere parish church. From this young man onwards our knowledge of the clockmaking activities of the Joyce family is well documented. William Joyce, his childhood behind him, emerged as a clockmaker at Wrexham in the neighbouring county of Denbigh, where he had no doubt been apprenticed to another member of his family, John William Joyce. What the exact relationship between the two was has not come to light—possibly uncle and nephew.

John William Joyce, clockmaker, was buried at Wrexham on 14 October 1717, and could have been the founder in 1690 of this ancient clockmaking firm.

The scanty evidence points to William Joyce succeeding John William Joyce in his Wrexham business, for in 1718 he is mentioned in the churchwarden's accounts of St. Giles, Wrexham: '1718. Pd Mr. Joyce for Cleaning the Clock and

Chimes & Keeping tham a going & for 2 pounds of brass wier & Cleaning ye watch in ye Church.   £2. 8s. 0d.'

He married his first wife, Lewcy Conway, at Wrexham on 27 February 1714/15, and it was there on 14 March 1715/16 that his first son, John, was baptised. It is sad to relate, but this boy (and presumably his wife) died, for in 1718 another son, John, is baptised; this time his wife's name is given as Mary. Two years later a daughter, Elizabeth, was baptised, and again his wife's name is given as Mary. He was still at Wrexham, plying his craft as a clockmaker, in the December of 1722, when he was named as the executor of his uncle, Arthur Joyce. The will reads: 'my nephew, William Joyce of Wrexham'. Arthur Joyce had been an Innholder of Cockshutt, and left all his real and personal estate to his nephew, William Joyce, which is perhaps the reason why he chose to leave Wrexham at this time to come back to his native place of Cockshutt, which he had done by the August of 1723, when his son, Arthur, was baptised at Ellesmere parish church. In 1724 he suffered a double bereavement; firstly his wife Mary, then barely a month later in the September, his year-old son Arthur. Still a relatively young man, he remarried for the third time within the year, to Ann Jones of Ellesmere.

On the decease of Edmund Bullock, clockmaker, in 1734, William Joyce took over the maintenance of the clock at St. Mary's church, Ellesmere. His salary, paid by the churchwardens, was four pounds per annum. Late in 1738 he passed this job on to his newly-married son, John, then aged twenty. William Joyce died in 1771, eight days after his eighty-first birthday. His widow, Ann, died a year later, within six days of the anniversary of his death.

John Joyce had worked at the trade of clockmaking with his father, William, and took over the family concern. As already mentioned he succeeded to the post of maintaining the clock of St. Mary's, Ellesmere, from his father in late 1738, and continued to look after it for 20 years (1738–1758). In 1747 he was paid 10 shillings for altering the dial, and in 1752 he received a further 10 shillings and sixpence to repair it. It is interesting to note that it was not below his dignity to attend to odd jobs about the church.

'1740. paid John Joyce for cutting the Brasses 15/-.
1747. pd John Joyce for mending ye key of ye Steeple
0. 0. 3.
1766. John Joyce for Cleaning the Candlestick Twice.
0. 10. 0.'

In 1738 John Joyce had married Deborah Sadler at St. Mary's, Ellesmere. When not occupied in making and repairing clocks and watches, John had his hands full caring for a large family, which consisted of seven sons and one daughter. Of the sons it has been possible to identify five who became clock and watchmakers. One son, Malachi, died in infancy in 1757, while William Joyce, born in 1748, remains an enigma; for apart from his baptism registered at Ellesmere, nothing more has come to light regarding him. The only daughter, Elizabeth, born in 1742, married in 1769 Thomas Jones of Whitchurch, a tanner by trade. John Joyce died on 1 November 1787, in his seventieth year. He was buried at Cockshutt; his sister, Elizabeth, lies in the same grave. The stone chest tomb erected to their memory is remarkable in that the memorial inscriptions are engraved on a brass dial intended originally for a broken arch long-case clock. There is a well-executed engraving of an angel sounding the last trump from a very fleecy-looking cloud within the arch area, below which is the inscription:

'UNDERNEATH
lieth the remains of John Joyce
He died the 1st Novm^r 1787 Aged 70
ALSO
Elizabeth Joyce Sister to the
above John Joyce
died 12. September 1792.
Aged 72.'

Deborah Joyce, John's widow appears to have gone to live with her eldest son, John, at Ruthin in the neighbouring county of Denbighshire, or died while on a visit to him, for her death is recorded in the registers of the parish church of Ruthin:

'1799. May 6. Deborah Joyce buried.'

To avoid confusing the reader, it will be best to follow the story of this family through the succession of generations working within the county, and to leave the stories of the other four sons of John Joyce, who left the county to set up their own businesses, until later in the chapter.

James Joyce, born in 1752, the third son of John and Deborah Joyce, took over the family business at Cockshutt. It was James who later decided to move the business from Cockshutt to Whitchurch. When exactly the move took place is not known, but he had moved by the year 1782. In that year, then aged 30, he married at Wem, Sarah, daughter of Benjamin Barnett of Soulton Hall. The entry in the register of St. Peter and St. Paul at Wem records James as of the parish of Whitchurch. It appears by the churchwardens' accounts of St. Alkmund's, Whitchurch, that James Joyce took over the maintenance of the church clock from Richard Deaves. The first entry referring to him is in 1784, when he paid 'by bill 3. 13. 0.' In 1806 he was still being paid for making repairs to the clock. Sidetracking for a moment in the accounts of St. Alkmund's, it is of passing interest to note the custom of those days for the parish officials to pay boys to kill vermin. Under the heading of vermin was listed the poor little sparrow, and three entries in 1788 list payment to 'Joyce's Boy' when he was paid for a total of 13 dozen sparrow-heads, at the rate of twopence per dozen.

In the March of 1790, James Joyce, styled 'Clockmaker of Whitchurch' decided to purchase the premises where he carried on his business. The owner was William Wicksteed of Whitchurch, the property:

> All that Messuage or Dwelling-house with the yard Stables & garden thereto belonging & appertaining situated & being in or near to a certain Street called the High Street now in the tenures or occupations of the sd James Joyce & Ann Cross widow And also all that other Messuage or Dwellinghouse with the yard stable Warehouse & gdn thereto belonging & appertaining situated & being in Whitchurch afsd in or near the sd Street called the High Street & now in the tenure or Occupation of Thomas Overton his Undertenants or Assigns.

This property was purchased by James Joyce for the sum of £660, no small sum in those far-off days. In 1795 he was elected churchwarden of St. Alkmund's, Whitchurch. The following advertisement appeared in the *Salopian Journal*, dated 3 May 1815:

> LOST. on the 7, April between Shrewsbury & Llandrinio. A SILVER WATCH, (Maker's Name JOYCE, Whitchurch) with a Gold Seal having the letter P with a crest above engraven upon it. Whoever will bring it to THE PRINTER OF THIS PAPER, shall receive ONE GUINEA Reward.

James Joyce, like his father before him, was very much the family man. Between 1783 and 1796 his wife, Sarah, presented him with nine children—five sons and four girls. Unfortunately, they lost two sons and one daughter in infancy. This left them with three sons and three daughters: Elizabeth 1783; Ann 1784; Sarah 1791; John Barnett 1792; Thomas 1793; and Richard Owen 1796. Of the sons, John Barnett Joyce joined the Royal Navy, becoming a lieutenant, dying at Preston Brockhurst in 1822, in his 29th year. Thomas Joyce followed in the family business; while the youngest son, Richard Owen Joyce (named after the great social reformer) remains an enigma, nothing being known of him beyond his baptism.

Sarah Joyce died in the year 1808, at the age of 54; the *Salopian Journal* briefly records her passing:

> 'Saturday last, Mrs. Joyce, wife of Mr. Joyce, watchmaker of Whitchurch.'

James Joyce died on 23 December 1817, aged 65.

Thomas Joyce followed his father, James, in the High Street business. He takes John Smith to court in the October of 1821; the entry on the Quarter Sessions Roll reads:

> John Smith of Whitchurch, labourer pretending to Thomas Joyce of the same, watchmaker, that he was a servant of Mary Congreve, spinster, a customer, and obtaining from him 1 dozen large knives and forks, 1 dozen small knives and forks, 1 carving knife and fork of steel. Fined £4 or 3 months solitary confinement.

Thomas Joyce married Charlotte Jones of Gresford, near Wrexham, resulting in a healthy family of six children—four sons and two daughters. Of the sons, James, *c.* 1821, became a watchmaker; John Barnett, *c.* 1826, also followed his father's trade: Conway, c. 1827, does not appear to have had anything to do with the family concern; Thomas, c. 1831, is returned on the 1851 Census as a farmer's apprentice, though still living at home.

In the *Pigot's Directory* of 1834, Thomas Joyce is shown as 'Watchmaker & Engraver', while in the Post Office *Directory* of 1856, he had branched out as 'Maltster and Agent to the Provident Life Assurance Office'.

In 1846, Archdeacon John Allen came to Prees as its vicar, and it was soon after that E. B. Denison (later to become Lord Grimthorpe) started what were to become regular visits to his friend. He was a personality who was to become the greatest authority in the world on all matters connected with clocks. So fond of clocks was he, that whenever he visited Archdeacon Allen he would drive over to Whitchurch, and spend the whole day in Thomas Joyce's workshop. After Lord Grimthorpe's quarrel with Dent's of London, he turned his attention to other clockmaking firms which were capable of working to his critical specifications. Thomas Joyce measured up to this great man's high criteria, and his clocks were recommended far and wide by E. B. Denison. An excellent example of the fine craftsmanship of Thomas Joyce is to be found in the church clock of St. Michael's, Marbury, in neighbouring Cheshire, made in 1844, as illustrated in this book. This clock was overhauled for the first time in 1955, after 111 years' service, and although the bearings needed re-bushing, nothing else appeared to have shown signs of wear, and it was declared to be good for another century.

Thomas Joyce was a progressive man, always ready to take an interest in and encourage new ventures. In 1859 he was presented with a solid silver salver, inscribed by the shareholders of the Whitchurch and Doddington Gas Company, as an acknowledgement

'of his valuable & Gratuitous services'

in establishing and promoting the prospects of their company.

On the Census Returns of 1851, Thomas Joyce is shown as employing two apprentices: Joseph Elliott, and Joseph Ridgway. It is of interest to note that 34 years later, in 1885, Joseph Ridgway still appears on the paysheets of the firm, while his 13-year-old son, Bruce Edward Ridgway, had joined him in 1884 as an apprentice.

In 1855 the Corporation of Shrewsbury decided that a new clock was badly needed for the old Market Hall in the Square. The erection of this was duly reported (at some great length) in the *Salopian Journal*. The report shows that Thomas Joyce had already acquired a high reputation for the standard of his work, and it is perhaps fitting to quote here an abridged version of the report:

> NEW CLOCK AT THE MARKET-HOUSE, SHREWSBURY.
>
> Last week the new clock, with an illuminated dial, ordered by the corporation for the front of the Market-House, was completed, and will, we have no doubt, prove of public advantage as the 'Shrewsbury Clock,' giving and keeping such true time as the inhabitants may wish to stand by. To secure this, the Town Council advertised for estimates, and that of Mr. Joyce, of Whitchurch, in this county, being the most eligible, was accepted, as it especially included all the latest and best improvements which mechanical and mathmatical science had suggested, combined with every practical application to ensure simplicity of construction with equality of motion; and so far as we have ascertained, this result has been affected in the most satisfactory manner.
>
> The new clock is what is ordinarily termed an eight-day one, requiring winding only once a week, and the hour will be announced on the old bell by a hammer more than double the weight of that formerly used, by which the sound will be made more effective.
>
> The dial is one piece of plate glass, four feet in diameter, having a ground of white enamel with black Roman numerals, and looks exceedingly well at night, when illuminated by four burners, which diffuse a full and clear light equally over its surface. The gas we understand, is liberally supplied gratuitously by the Gas Company, and will be kept always slightly burning, but the working of the clock is so made to turn the light nearly off, or full at the proper times, by a twenty-four hour wheel, with pins set in by hand, as the length of the day varies.
>
> Altogether the mechanism of the clock appears to have been got up in a painstaking and excellent manner, and we trust that by the precision with which it will perform its deliberate purpose,—to indicate the fleeting hours of time,—it will prove creditable to the experience and reputation of Mr. Joyce, the maker, as well as useful to the town and neighbourhood generally, Hence 'tenders' for a clock of the best construction, and containing all approved requisites for accurate performance were advertised for; and the tender sent in by Mr. Joyce, of Whitchurch, was accepted. As Mr.

Joyce (according to a treatise on *Clock and Watchwork,* just published) has long enjoyed the reputation of one of the best provincial clock-makers, we sincerely hope the resolution of our corporation will still go farther to establish him in the well-earned reputation he is now enjoying.

The construction of the new clock, with the exception of striking the quarters, and the substituting gun-metal wheels instead of cast-iron, is on the same principle as the large one made by E. J. Dent for the Parliament-House, viz. to go eight days, to strike the hours, to continue going while winding, patent iron-wire ropes, improved arrangement of the wheel work and frames, doing away with the old cage-pattern appearance, and substituting a method by which one wheel or part may be removed without deranging the whole, improved compound iron and zinc compensation pendulum rod, and last but not least, the newly invented 'gravitation escapement'. The last named improvement deserves more than passing remark. A short time since, a new clock was required for the cathedral of Fredericton, Norway, where the cold of winter descends to more than 40 degrees below zero; to counteract the great diversity of the amount of friction caused by such intense cold congealing the oil, Mr. Denison (under whose supervision the great clock for the Parliament-House was made) was led to invent this new escapement to meet the difficulty; it having the peculiarity of imparting an equal impulse to the pendulum under any amount of force that may be applied to the movement by more or less weight, the action of wind or weather on the hands, or any obstruction that may arise in the train from unequal friction in its working, provided the obstruction is not sufficient to stop it completely. It consists of two bent pieces of metal suspended on small pivots, and resting one each side of the upper part of the pendulum rod, which from its position prevents the pieces of metal falling into the line of the centre of gravity, which they would otherwise obtain; this causes a slight push on the pendulum rod; by a peculiarly formed wheel, consisting of three long teeth and three pins, these pieces are alternately lifted from, and let fall on the side of the rod, and though the pendulum ball weighs 200 lbs., yet the small force of those bent pieces alternately on each side with a power of less than half an ounce, keeps it in constant motion.

On a small dial, in the interior of the clock, is engraved the following inscription—

> 'William Butler Lloyd, Esquire. Mayor.
> Joshua John Peele, Esquire. Town Clerk.
> Mr. Henry Pidgeon. Treasurer. A.D. 1855.'

The style and finish of the workmanship reflect great credit on Mr. Joyce, but not more than the openness of mind with which he receives improvements and abandons old notions which have nothing but their antiquity to recommend them, but which are so tenaciously adhered to by the majority of clock-makers.

In the Census of 1861, James Joyce, watch and clockmaker, aged 40, High Street, and employing two men and two boys, is returned. His mother, Charlotte, now aged 70, is living with him. James was born in 1821, and is described in the various directories, 1856-1875, as watch and clockmaker, silversmith, jeweller and cutler. He remained a bachelor all his life, and died in 1883.

John Barnett Joyce, born 1826, a younger brother of James, appears to have spent a number of years at Bradford, in Yorkshire, where most of his children were born. By 1871 he is back at Whitchurch; the Census for that year gives his address as St. John Street, where as a turret clock maker, he employed six men and two boys. At this time his brother, James, is still at the High Street works, where he is shown as employing two men and two boys, which rather points to the fact that they were two separate establishments, whether in opposition or working in conjunction is not shown. But it is evident that John Barnett Joyce must have taken over his brother James's workshops on his death, if not sooner. The firm was world-famous by this time; Joyce clocks were fitted in nine cathedrals in this country alone: Bangor, Chester, Chichester, Lichfield, Christ Church, Oxford, St. Davids, Salisbury, Southwell Minster, and Worcester. Churches innumerable, universities, public buildings, railway stations—the output has been tremendous, but the quality has always remained as high.

It was while under the direction of John Barnett Joyce that the firm first became known as 'J. B. Joyce & Co.', a title which it has kept to date.

It was a great honour when Joyce's were commissioned to make a clock for Salisbury Cathedral in 1884, although it was not known at that time how ancient the original clock, then still in use, was. Mentioned in 1386, it was the oldest clock (*in situ*) still working. In 1870 the Corporation of Shrewsbury decided to purchase another Joyce clock, this time for the new Market Hall, which faced on to Shoplatch.

The frame of the clock is horizontal, cast in one piece. The pivot holes are screwed on to the frame so that each wheel can be separately removed, when the clock requires cleaning, without disturbing the frame. The whole of the wheels are of gun-metal, and the pinions lantern ones of hardened steel.

The maintaining power is the going ratchet. The principal feature in the clock is the gravity escapement, which is now proved to be the best to secure accuracy in the going of a turret clock, as it renders the pendulum independant of all variations of force and friction in the clock and sufficient weight can be attached to drive the hands of four large dials in all weathers, without influencing the vibration of the pendulum. The pendulum is secured against the effect of variation of temperature, being compensated with tubes of zinc and iron. It weighs about two hundredweight; and is so attached to the frame as not to require removing to clean.

The striking part of the clock is very powerful, and lifts a hammer 60 lbs. weight. The dials are in one sheet of place glass 8ft. 9ins in diameter, and, indeed, are supposed to be the largest that have yet been made in one piece. Unfortunately, two of the first four dials were broken, one in carriage, and one in getting it up the tower. Two new dials having to be made, and it being necessary both dials and bell should be on before the clock could be fixed, caused considerable delay. The two new dials were differently figured, by order of the committee, from those originally made. The clock is placed 50 feet below the dials. There is machinery attached to the dial work for turning on and off the gas. The weight of the going part is 60 lbs., that of the striking nearly 4 cwt.

On the plate on the regulating dial in the interior is engraved the following:—

'JOYCE, WHITCHURCH. 1870.

This clock was erected by public subscription, chiefly promoted by H. Keate. J.P., chairman of the Market Committee; Henry Fenton, Mayor; E. Cresswell Peele, Town Clerk; Henry Pidgeon, Treasurer; Thomas Tisdale, Surveyor; Robert Griffith, Architect.'

Arthur Joyce, born c. 1863, at Bradford, son of John Barnett Joyce and his wife, Ellen Roberta Joyce, joined his father's business, eventually taking over its management. A very fine long-case clock, its brass dial inscribed: 'A. JOYCE. WHITCHURCH.' in a carved oak case, is still in the possession of the family. Arthur carried on the fine reputation of the firm, sending the products of his workshop all over the world:

Australia, Burma, Canada, China, Egypt, Falkland Islands, Gambia, Gibraltar, India, Malay Straits, New Zealand, Nigeria, South Africa, and South America. The Right Hon. Lord Grimthorpe, writing from his home, Batchwood, St. Albans, on 9 December, 1902, says:

> Messrs. Joyce, of Whitchurch, are one of the only three firms I have been in the habit of recommending as competitors for contracts for large clocks, stipulated not to differ from Greenwich time by telegraph more than two seconds in a week. I have also inspected a good many of their clocks while I was able to travel to see them. My turret clock here, made by them, still goes admirably; it is 30 years old this summer, I think.

Lord Grimthorpe (the greatest authority of his day on clocks) never stinted in his praise of this fine firm. The great clock which the firm sent out and fixed in the tower of Sydney Post Office, Australia, was made under the supervision of Lord Grimthorpe, who declared it to be one of the finest pieces of clockwork he had ever seen or it was possible to make. Later, when asked by one of the Colonial governments to give his opinion of clock-makers, he declared Messrs. Joyce to be the greatest makers of clocks in the Empire. In 1904 new premises were erected in Station Road, close to the railway station, and fitted with modern equipment for turret clock making. Still open to new ideas they fitted their own electrical generating plant, and lit their new premises with their own electricity. Another innovation: they were the second to introduce the motor car to Whitchurch. This was in 1904 and the number plate read: 'AW 56'. Two years later in 1906, when the Whitchurch telephone exchange opened, the firm soon grasped the advantage of rapid communication, and from then on their stationery bore their telephone number, 'Whitchurch 17'.

Arthur Joyce had at first been in partnership with his brother, Walter Joyce, but after his death, became the sole head of the business for some 12 years. His wife, Jessie Powell, bore him six children: Dorothy, Jack, Norman, Arthur, and William. Of the boys, Jack Joyce, a one-time member of the Royal Canadian Mounted Police, worked in the firm for a short period after the First World War, but had little interest in clockmaking. William Joyce emigrated to Australia, like his brother Jack, not interested in the family business. Arthur Joyce was unfortunately killed in France, December 1917. This left only Norman Joyce, born 1891, to carry on the family concern, when his father died at the early age of 49, in 1911.

In 1915 this firm constructed a turret clock for the tower of the Post Office building in the city of Lindsay, Canada.

When the building was demolished a few years ago, the city council bought the clock, with the idea of having a tower built to contain it, but the cost of building such a tower specially for this purpose proved to be more expensive than they had anticipated, with the result that the clock was kept in store.

However, Robert Phillip, owner and curator of the 'Museum of Time' which stands just outside Cookstown, Canada, learnt of it, and eagerly purchased it from the city fathers. A tower was erected on the corner of one of the outbuildings of his museum, and the clock installed. Originally the clock had four dials, each 3ft. 6in. in diameter, which enabled it to be read from any direction, but, sad to relate, one dial and its gear train were badly damaged in the demolition of the original Post Office tower, so that now the clock only shows three faces. It is nice to think that this Shropshire-made clock is now preserved for posterity in its museum home, some 62 years after its making.

Under the skilful guidance of Norman Joyce the firm continued to flourish, sending clocks all over the world—one of the outstanding triumphs was the clock for the Shanghai Customs House, which he went out to China to fit himself—an enormous clock, the dials were 18ft. in diameter, weighing 1½ tons each. Minute hands 10½ft. long. Weight of each pair of hands 2cwts. Pendulum 15ft. long, weight 5cwts. Main wheels 1cwt. each. Hour bell 6¼ tons. Four quarter bells 4 tons. Hour hammer 280lbs.

Extract from the *North China Daily News*, 2 February 1928:

'SHANGHAI CUSTOM HOUSE CLOCK'—TIMEKEEPING EXTRAORDINARY'

The movement of 'Big Ching' has now been adjusted to minute accuracy, and for the past fourteen days the clock has not varied so much as one second from the Time-ball on the French Bund.

> Shanghai has experienced very trying weather in the past fortnight, and such a record for a turret clock, whose hands are exposed to all the winds that blow, is an achievement of which the makers may well be proud. The community, both Chinese and foreign, is finding this standard time which is now available in Shanghai and neighbourhood, a great boon and the wireless broadcasting of 'Big Ching' ensures that the outports also receive accurate zone time daily.'

Railway clocks, large and small, including the well-known pattern of wooden-cased station-room clocks, form a large feature of the firm's business. In 1909 they advertise that they have—

> 'supplied, keep in order and wind up, in large centres such as Liverpool, Manchester and Crewe, some 1,500 clocks of various sizes on the London and North Western Railway alone, in addition to hundreds for the Lancashire and Yorkshire, the Great Western, and most other English and Foreign railway companies.'

Sad to relate, Norman Joyce was destined to be the last of his family to carry on this ancient craft of clockmaking at Whitchurch. With no sons to carry on after him, he decided in 1965 to retire, and the firm merged with that of John Smith and Sons, of Derby, another very old-established clockmaking firm. But J. B. Joyce & Co., Ltd., still retain their name and individuality, their men still travelling all over the country, erecting new, and cleaning the old turret clocks of Britain.

Just after Christmas 1966, Mr. Norman Joyce died suddenly, at the age of 75, after a lifetime in the service of his family concern.

Eight generations of clockmakers had come to an end.

* * * * *

To go back in time, to the four sons of John Joyce (1718-1787), who left the county to seek their fortunes elsewhere:

Robert Joyce, son of John and Deborah Joyce, was baptised at Ellesmere on 20 June 1754. This young man left the county to serve his apprenticeship in London, after the expiration of which he served with several clockmakers of note in that city, before leaving for Dublin where he set up in business for himself for a period of seven years. He is recorded as a witness at a wedding at Wem in 1774, with his brother, James, probably on a visit home. He left Dublin for America, and set up in business in 1794 at No. 4 Beaver Street, New York:

> ROBERT JOYCE.—Watch & Clock-Maker, No. 4, Beaver Street, Takes the liberty of informing his friends and the public, that he has commenced business in this city. Having served his apprenticeship in London, and afterwards wrought with the most eminent in his line there, and in Dublin; and in the latter place for seven years carried on business on his own account, in the course of which he has been employed in making time pieces for astronomical observations, Airometers for shewing the point of the wind, and the Clocks for the principal part of all the public buildings; so that his experience gives him confidence to assert, that with his strict attention, he will execute every command in his line of business in the best and most satisfactory manner.
>
> Wishing to be respectable, Joyce will not undertake the repairing or cleaning of any Watch or Clock, without being first permitted by the customer to make good the defective parts, which by having every necessary engine and tool for the purpose, he can do (however intricate the work) in a more reasonable and satisfactory manner than by half doing and often charging.
>
> He has at present a variety of Clocks, which he will engage to the purchaser; also Gold, Silver, Enameled and Mettle Watches, of his own make, from two Guineas to fifty, with gold hands, keys, glasses, and every other article in his line of business of the best and cheapest kind.

The above advertisement appeared in *The Diary; or Evening Register,* January 8, 1794.

He appears to move very frequently about the city of New York, for in 1794 his place of business is: No. 4 Beaver Street; in 1795, No. 32 Beaver Street; in 1796, 62 Wall Street; 1797, at 149 Pearl Street; then in 1798 (the year of his death), at 145 Pearl Street. In 1797 he inserts an advertisement in the *New-York Daily Advertiser,* dated May 22, 1797:

> ROBERT JOYCE.—Sign of the Eagle and Watch, next house to the corner of Wall street, in Pearl street, nearly opposite the Coffee house slip. Robert Joyce, Watch and Clock-maker, having removed to the above situation, respectfully inform his friends and the public in general that he has considerably increased his Stock of Watches. He has a variety in gold, silver and metal cases; some in the English stile, and of the most superior workmanship; horizontal, capped and jewelled, in uncommon strong gold cases, made under his own immediate inspection; and having been regularly instructed

under the most eminent in his line, in London, he with confidence asserts, that they cannot be excelled any where; and hopes, by his unremitting attention, and desire to perfect every piece of Mechanism that passes through his hands, to merit a continuance of that liberal encouragement which he has so amply experienced since his commencement in this city.

Such watches and clocks as he has occasion to import, are described according to his own ideas, and he trusts more for the advantage of his customers, than those imported by persons ignorant to the true principles and execution of such machine. Some of Litherland's patent watches with jewelled pallets and holes, marble and other clocks ready for sale. Also a variety of second-hand watches. Every description of watches and clocks cleaned and repaired; if ever so intricate, he will engage to perfect them. Any person wanting gold cases, can be immediately supplied with any pattern or strength. All kinds of gold bought.

He appears to have died in 1798, for after that year he fails to appear in the *New York Directories,* and instead we see:

1799. Widow of Robert Joyce, Pell Street.
1800. Widow of Robert Joyce, dry goods, 393, Pearl Street.
1801. Widow Eleanor Joyce, 6, Chapel Street.
1803. Widow Eleanor Joyce, 30, Chapel Street.

After which there is no mention of his widow. The Letters of Administration on his death were granted on 4 December 1798 to his friend, Joshua Sutcliff, of Philadelphia. That he had a son we do know, for he returned to England to serve his apprenticeship with his uncle, Samuel Joyce, of London:

Conway Joyce, jun; son of Robert, of New York, North America, watchmaker, deceased, was on January 7, 1811, bound apprentice to Samuel Joyce, watchmaker of 38, Lombard Street, London, for seven years, at a consideration of one penny.

There is no record of him as having taken up his freedom of the City of London on the expiration of his term. Presumably Conway Joyce went back to America, and it is of interest to note that there is a Thomas F. Joyce, a watchmaker in Philadelphia, 1820-25, the city where the executor of his father's will resided; one is left to speculate on the relationship.

Two more of the sons of John and Deborah Joyce were Samuel Joyce, baptised 28 January 1759, and Conway Joyce, baptised 5 April 1761. These two brothers acted in partnership and travelled to London where they laid the foundations of a prosperous business. They were active as a watch and clockmaking firm at Three Kings Court, Lombard Street from 1785 until 1789, and at 38 Lombard Street from 1790 until 1839. They traded under several styles: Joyce and Conway 1785; S. and C. Joyce, 1789-1839; and Samuel Joyce, 1840-1842. A fine gold engraved musical watch by these makers is to be found in the Feill collection, and a gold eight-day watch in the Denham collection.

Both Samuel and Conway Joyce became free of the Clockmakers' Company by redemption on 5 February 1810.

In the registers of the church of St. Edmund the King and Martyr are to be found the entries of baptisms and burials of the children of Samuel Joyce, junior:

> 3. Aug. 1834 bapt. Samuel Joyce son of Samuel and Bridget Ann.
> 21. Dec. 1834 bapt. Conway Joyce son of Samuel and Bridget Ann.
> 11. Dec. 1836 bapt. Thomas Joyce son of Samuel and Bridget Ann.
> 4. Oct. 1838 bapt. Anne Joyce dau of Samuel and Bridgett Ann.
> 1. Aug. 1841 bapt. William Joyce son of Samuel and Bridget Ann.
> 8. Jan. 1837 bur. Conway Joyce aged 2 years 2 months in churchyard.
> 14. Nov. 1841 bur. William Joyce aged 1 year 6 months in churchyard.

Samuel Joyce, senior, had died in 1827, and was buried in the churchyard of St. Edmund's on 13 September 1827, aged seventy. His brother, Conway, was buried in the same place on 6 May 1836, aged seventy-five. Samuel, son of Samuel Joyce, senior, then took over the business, the last mention of which appears in 1842.

Samuel Joyce, junior, served his apprenticeship to his uncle, Conway Joyce:

> Samuel, son of Samuel Joyce, of Lombard Street, London, clock and watchmaker, was on April 2, 1821, bound apprentice to Conway Joyce, also of Lombard Street, watch and clockmaker, for seven years at a consideration of 5s. He obtained his freedom by servitude on October 10, 1836.

In 1931 an interesting letter was received by Mr. Norman Joyce, of Whitchurch, from a distant relative, a Miss Annie Joyce, of Thornton Heath, Surrey. In it Miss Joyce relates:

I remember my father speaking of his great uncle telling him about George IV coming into the old shop in Lombard street, to see the beautiful jewelled watch which the firm had made for The Emperor of China.

John Joyce of Denbigh and Ruthin

John, the eldest son of John and Deborah Joyce, was baptised at Ellesmere on 1 June 1744. He, too, left home to seek his fortune, and in his case turned his face towards the county of Denbighshire, where his grandfather, William Joyce, had served his apprenticeship.

John Joyce was to become a much married man, for in his 65 years of life he had four wives and 13 children. His first wife was named Peregrina, but little is known of her except that she gave him two children—Lucy Conwy Joyce in 1768, and John Joyce in 1770. When she died he married his second wife, Elizabeth, and they also had two children: John Joyce in 1772, and Ann in 1773. It would appear that the first son, John, had died, for a second son was named John; however, the second John died also in 1773. The burial of the second son was at Denbigh: '5th January, 1773. John son of Mr. Joyce, Watchmaker buried'. His second marriage was also short-lived, for Elizabeth was buried at Denbigh on 6 February 1777. John did not let the grass grow under his feet; with two small children and a business to look after he was in dire need of help. On 14 May in the same year he married for a third time: '1777. May 14. John Joyce of Denbigh, widower, and Mary Courter of Ruthin, widow, were married by licence'. The wedding took place at Ruthin, and it was there that John and his bride settled down to live. There was no issue from this marriage, which lasted five years. Mary Joyce died, and was buried at Ruthin, 7 December 1782. In the November of the following year John entered the matrimonial stakes for the fourth time: '1783, November. John Joyce of Ruthin, widower, and Barbara Griffith of the same parish, spinster, were married'. After three short-lived marriages, this one was destined to last 26 years, during which time Barbara lovingly presented him with nine children: John 1784; Peregrina 1786; Robert 1788;

Elizabeth 1791, bur. 1792; Samuel 1793; James Griffith 1798; Thomas 1801; and Sara Maria 1804, bur. 1804.

Also in the parish registers of Ruthin is to be found the entry of his mother's burial: '1799, May 6. Deborah Joyce buried'. It is possible that she had died while on a visit to him, or even that she had gone to reside with him after her husband's death. John Joyce was to die in 1809; his wife Barbara remained a widow for 19 years, dying in 1828.

Of the sons of John Joyce, three became watchmakers: John Joyce II (1784-1847), Robert Joyce (1788-1859), and James Griffith Joyce (1798-1874). It would appear from family wills and papers that John Joyce II took over the family business from his father, but later took his brother, Robert, as a partner. They are shown in local directories for the period 1835-1844 as 'John & Robert Joyce, Watchmakers, Well Street, Ruthin'. A long-case clock (No. 39) with a white painted dial, inscribed 'J & R. JOYCE' was formerly in the possession of Daniel Owen, the Welsh novelist (1836-1895), and is now in the keeping of the Welsh Folk Museum.

John Joyce married in 1828 Catherine Jones; a childless marriage. Apart from his watchmaking activities, he is described in Slaters's 1844 *Directory* as 'sub-distributor of stamps, Well Street'. All three brothers are listed as Freemen of the Borough of Ruthin on an Electoral Register dated 1836.

John Joyce II died in the year 1847, and his wife 10 years later in 1858.

Robert Joyce, as the surviving partner in the family firm, carried on the watchmaking business in Well Street. Britten mentions a watch by this maker 'Watch 1819' by 'Robert Joyce'. An interesting sidelight on the family occurs in an advertisement in *Eddowes Journal*, dated 23 February 1859, under the heading: 'The North And South Wales Bank.—A List of the Persons of whom the Company or Partnership consists. Amongst a great many other names is: 'Robert Joyce, Ruthin, gentleman', and also 'JOYCE executors of John. Ruthin. watchmaker'.

Robert Joyce and his wife, Ann had four sons: John, c. 1831; Robert Griffith, c. 1828; Henry, c. 1837; and Thomas. Of these, John became a publican, and was with his wife, Mary, licensee of the *Boar's Head* inn, Clywd Street, Ruthin, appearing as such on the Census Return of 1861. Thomas Joyce emigrated to Australia, where in the fullness of time he died. Both Robert Griffith Joyce and his brother, Henry Joyce, were apprenticed to their father's trade of watchmaking.

Robert Joyce died in the year 1859, his business in Well Street was inherited by his son, Robert Griffith Joyce.

To go back in time to the third son of John Joyce I: James Griffith Joyce, born 1798. This maker appears in local directories from 1835 until 1868 as 'James Joyce, watchmaker, Clywd Street, Ruthin', and in the latter year with the addition of '& Son'. He married Charlotte Price, born in 1800, at Oswestry, who presented him with three sons, and one daughter: Thomas, c. 1828; James, c. 1830; Charlotte, c. 1831; and Walter Conwy, c. 1840. James Griffith Joyce died in the year 1874; his business was carried on by his son, Walter Conwy Joyce.

In 1874, Worrall's *Directory of North Wales* describes Walter Conwy Joyce as 'Walter. C. Joyce, Watchmaker & Jeweller, Authorised Agent to the Singer Sewing Machine Company, 1, Clywd Street'. He died on 14 November 1910.

Returning to Robert Griffith Joyce, who had inherited his father's business (Robert Joyce, of Well Street), he appears in local directories from 1856 until as late as 1887 as 'Robert. G. Joyce, Watchmaker, Upper Well Street'. In Worrall's *Directory* of 1874; the address is more specifically given as '4, Well Street'.

A watchpaper issued by this maker is in the collection at the Welsh Folk Museum.

The third son of Robert Joyce, Henry Joyce (c. 1837-1909), also became a watchmaker. In the Census of 1861 he is returned under 'Llanrhydd, Wernfechan':

> Henry Joyce, married, aged 24, Watchmaker, born Llanrhydd.
> Elizabeth Joyce, wife, age 26, born Corwen.
> Elizabeth. A. Joyce, daughter, age 1, born Llanrhydd.

Henry Joyce married Elizabeth Pearce on 5 September 1858, and as has been seen, he was at Llanrhydd in 1861, but removed his business to Denbigh, c. 1864. John Joyce, his grandfather, removed the business from Denbigh to Ruthin in 1777, while Henry Joyce brought it back c. 1864, nearly a century later.

Henry Joyce is returned on the Census of 1871 at Denbigh at '2, Vale Street' with a family now grown to two sons and three daughters, while in his shop he employed as an apprentice Richard Evans of Welshpool, aged fifteen.

The Welsh Folk Museum has four watchpapers of this maker. He soon moved from 2 Vale Street, for Worrall's *Directory* of 1874 puts his address as 13 Vale Street. Other directories from 1868 until 1890 list him as 'Henry Joyce, Watchmaker, Vale Street, Denbigh'. Henry Joyce died on 5 September 1909.

Of the three sons of Henry and Elizabeth Joyce, Henry Edward, born c. 1864, became a watchmaker, but it was John Parry Joyce, born 1872, who carried on his father's business at Vale Street. His marriage to Mary Kate Astbury in 1898 resulted in eight children, of which his son, John Trevor Griffith Joyce, born 1903, was destined to be his successor in his watchmaking business. John Parry Joyce died in 1948.

John Trevor Griffith Joyce kept this old-established watch-making business flourishing in the best traditions of the craft, and was well known and respected for his workmanship. When he died in 1971 the business was sold, but still carries on under the name of Joyce, a reminder of the long association of that family with Denbigh. So we still have a watchmaker's shop at Denbigh, Clwyd, and a turret clock-making firm at Whitchurch, Salop, both under the name of Joyce, but, unfortunately without a member of the family carrying it on. There were eight generations, from which stemmed 25 clock and watchmakers, spanning the reigns of 14 monarchs of this country.

J. B. Joyce & Co., Ltd., Whitchurch 157

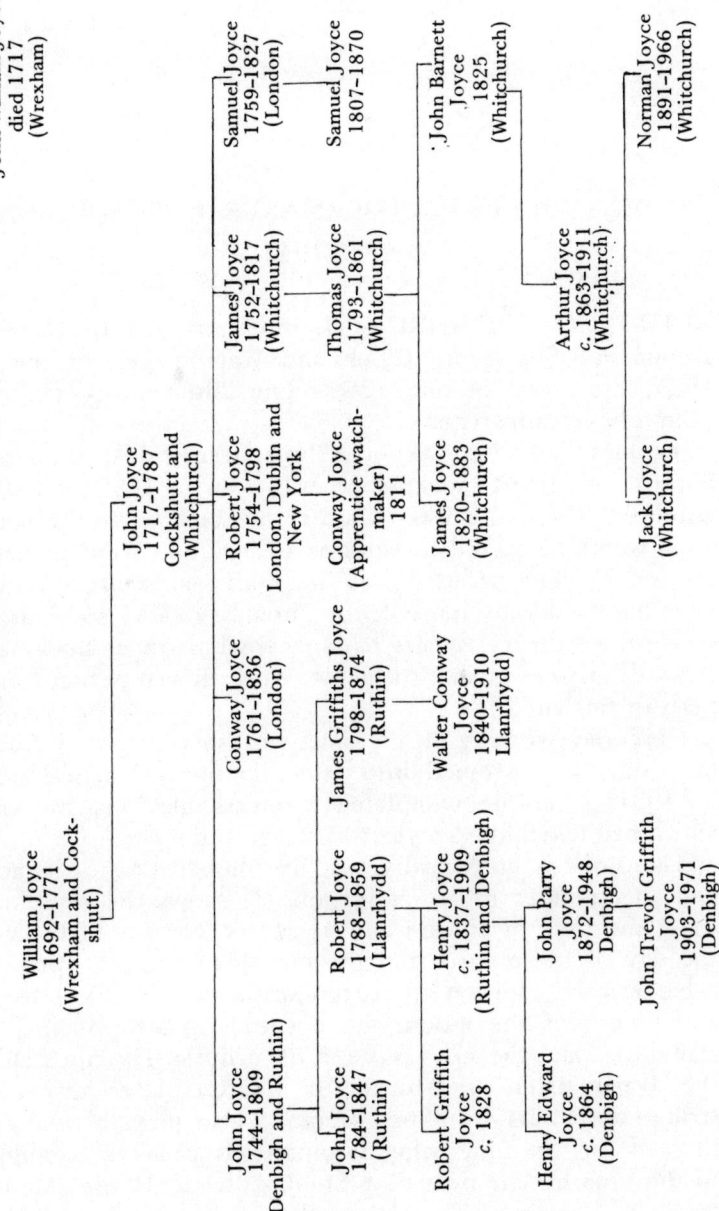

THOMAS VICKERY, CLOCKMAKER EXTRAORDINARY
1882–1945

ALTHOUGH THE ORIGINAL intention was to close this account of Shropshire Clock and Watchmakers at the year 1875, the story of one outstanding 20th-century craftsman cannot be left unwritten.

Thomas Vickery, of Astley Abbots, near Bridgnorth, possibly Shropshire's finest clockmaker, was born in 1882 at Bath in Somerset. From the West Country he came to the Bridgnorth area, where he stayed to work as a Clock and Watchmaker for the last 35 years of his life. A fine craftsman, most of his tools were hand-made by himself, for Thomas Vickery specialised in out-of-the-ordinary repairs to clocks, which most clockmakers refused, or were unable to tackle. Work flowed to him from all parts of England.

Obscurely working in his small workshop at Astley Abbots, he suddenly blossomed into fame (in the horological world) in 1931, when he completed a remarkable long-case clock, which had taken him 25 years to design and make.

The clock is noteworthy for the completeness of its action and the number of its indications. It shows the equation of time, the time of sunrise and of sunset, the day of the week, the day of the month, and the month of the year (perpetual calendar). In addition its astronomical dial indicates the age and phases of the moon, the position of the principal constallations, and the sun's place in the ecliptic. The clock chimes the Westminster or Whittington quarters alternatively and strikes the hours from one train. It also plays a tune every three hours, the tune being automatically changed at midnight to the tune for the next day: Sunday, 'Easter Hymn'; Monday, 'Stella'; Tuesday, 'The Harp that once in Tara's Halls';

Thomas Vickery, Clockmaker Extraordinary

Wednesday, 'All Saints'; Thursday, 'Ye Banks and Braes'; Friday, 'Come all ye Faithful'; Saturday, 'Tom Bowling'.

The small dials inside the numerals show the day of the week on the left one and the day of the month on the right dial, while the slot between them shows the month. Months of 30 days and 29 February in leap year are shown automatically by the right-hand dial finger. The scale underneath the figure XII on the chapter ring depicts what astronomers and navigators term the equation of time—the variation between Greenwich mean time and solar time. The clock has a Graham dead-beat escapement and a seconds pendulum with invar rod and has been designed and produced entirely by Thomas Vickery.

Before commencing to make this fine clock Thomas Vickery carried out the calculations himself, and made a diagram of every detail in itself a formidable task. The gear-cutting and wheel-making (one wheel has 487 teeth) and pinions are all his own handiwork. The fine case of pollard oak has also been made by him. After three years the clock had only lost 3.4 seconds. First shown at a local exhibition at Bridgnorth in 1932, it was later shown at the 1933 Goldsmiths' Conference at Llandudno. Perhaps the greatest attention was paid to this remarkable clock when it was exhibited at the 'Arts and Crafts Exhibition', at Dorland House, London. Again it went on show at the 'Birmingham Scientific Institute Conversazione' in 1936.

In an interview in 1935, Thomas Vickery was asked, 'What was your reason for making the clock?', and he replied: 'I have done it simply as a spare time hobby in order to solve mechanical and astronomical problems'. Then, after a pause, he added, 'And partly to leave something behind which was made in the twentieth century. There is very little made today which is worth leaving behind'.

This noteworthy clock now stands in the Clockmaker's Museum at the Guildhall Museum in London, the gift of Thomas Vickery.

This talented craftsman also made a remarkable planetarium, which was intended to be the first unit of another long-case clock. In the centre is a disc depicting the sun, and discs representing the principal planets which move anti-clockwise round the 'sun' in the time that is actually taken: Mercury 87 days,

Venus 224 days; Earth 365 days; Mars nearly two years; Jupiter 12 years; Saturn 29 years; Uranus 84 years, and Neptune 164 years.

At the same time a disc representing the moon moves around the earth, while the outer silvered ring is moving at such a rate that it will take 2,586 years to travel completely round. Thomas Vickery estimated that it would take 300 years before any error in the position of the planets would become noticeable.

Thomas Vickery died in 1945, and was buried in the churchyard of St. Calixtus, Astley Abbots. His stone reads:

TOM VICKERY,

OF ASTLEY ABBOTS,

CLOCKMAKER & TALENTED CRAFTSMAN,

CALLED TO REST DEC. 2nd 1945.

AGED 63 YEARS.

APPENDIX

Although not relating to a Shropshire clockmaker, it has been thought of interest to include a series of letters written by John Briant, a notable clockmaker of Hertford, who made a clock for St. Julian's church, Shrewsbury, and at the same time one for what is now the Castlegates branch of the Salop County Library, the latter still being in use.

(1)

To the Vicar, ChurchWardens, and Inhabitants of St. Julian's, Shrewsbury.

Gentlemen,
    I propose making a new three part Church Clock on a large Scale.

The Hour and Quarter part of 30 hours continuance, the Watch part of 8 Days continuance, The Wheels of fine Brafs, the Pinions and Pivots of Steel. One Convex Copper Dial between 5 and 6 feet Diam^{er} with hour and Minute hand neatly Painted and gilt, and Machinery to carry the Hands.

The Whole of the before mention'd businefs I will execute and fix in a sound and Workmanlike manner, on the most approv'd of simple construction, highly finish'd in every part, for the Sum of £145. Or, I will make a two part Clock on the same Scale, Workmanship, & Dial etc., as before described, for the sum of £95. The Money to be paid at any future time that you shall think necefsary for a Trial of the performance of the Clock, an that you are fully satisfied that I have executed the Businefs with fidelity, and according to

                      I am Gentlemen
                      Your obedient Servant
                      John Briant
                      Hertford
                      28th May 1812.

    N.B.—This Estimate is calculated to receive every Afsistance of a Carpenter, to make a Case for Clock and fix the Dial.

(2)

ADDRESSED: The Churchwardens, St. Julian's, Shrewsbury.

GENTLEMEN,

The old Clock at the School is made upon too small a Scale, to do its Office in the present situation. I propose making a new 8 Day Clock upon a much larger Scale, the Wheels of fine Brafs, Pinions, Pivots, and every other part of action of Steel, the whole will execute on the most approv'd of simple construction, highly finish'd in every part and fix'd compleat exclusive of the Afsistance of a Carpenter to make a Case &c. for the sum of £100.

> I am your obedient Servant
> John Briant
> Clock-maker
> Hertford
> 25 th June 1812.

(3)

Revnd Sir,

I will work for my Shrewsbury Friends upon any Terms, and will reduce my Estimate for St. Julians to the Terms you conceive to be right of the School Clock, 'tho I afsure you there is no Inconsistency in my Estimate, the two Clocks are very different in Value, St. Julians must be upon a much larger Scale, consequently more Materials and much more workmanship. The conversation that past between Mr. Rowland and me, was, that a 2^r movement added more than 1/3 part to the workmanship and materials. Mr. Rowland suppos'd that 30£ I ansrd, Thereabouts. The following calculation I hope will meet your approbation.

	£
Estimate for School Clock	100
A 2^r movement added	30
	£130

I will make a 2^r Clock for St. Julians for £130 hoping I shall fix both at the same time, this and some other considerations induce me to make this Clock 20 p cent cheaper than I ever make a 2^r Clock before. Afsuring you tho it is too low a Price it shall have no effect on me in slighting the Workmanship, but I will use my utmost indeavour to merit your future

I have offer'd more than the value for the old materials, the School Clock I have some chance of fixing it, after much reparation.
St. Julian's is not worth 5£ to no person.

      I am Res^d, Sir
       your obedient servant
        John Briant
         Hertford 13th July
          1812.

(4)

       Shrewsbury 23. Sept. 1812
Sir

In consequence of you reduceing your Estimate for a three part church Clock on a large Scale for St. Julians we wish you to Make one with all convenient speed, the quarters to strike on the forth & 3 Bell & the Hour on the large do. We wish the Dyal Round or Octagon black ground with guilt Edge & Letters if you recollect I told you we had no chance to obtaining leave at a Vestry meeting for a new church Clock our Loans being so high we are doing it on our own Heads with the Afsistance of the Worthy Rev. H. Owen by Subscription & have the pleasure to Inform you we have Names for about one Half . . . hoping for our Credit as well as your Own. Use every Exertion in your Power to make in every part Good & Durable which will very much Oblige
      Your Hum^{ble} Serv^{ts}.
       R. Williamson
       J. Peele.   (SBL)

John Bryant (Briant), the bell-founder and clock-maker, was born at Exning, in Suffolk. He was intended to take up holy orders, but his love of mechanism was so strong that he was allowed to follow his natural bent. He soon acquired a pre-eminence in the trades of bell-founding and clock-making which has rarely been attained by any other individual. His clocks were widely sought and amongst his patrons were the Dukes of Marlborough, Rutland, and Grafton; the Marquises of Exeter and Salisbury; the Earls of Hardwicke, and Cowper; the Lords Montague and Breadalbane. The fine clock on the Town Hall of Hertford is an example of his work. John Bryant died in 1829, aged 81, and is buried at Hertford. His bells and clocks in the neighbourhood of Hertford bear various dates between the years 1782 and 1824.

LIST OF SHROPSHIRE CLOCK AND WATCHMAKERS BY TOWNS AND VILLAGES

Albrighton
Baddeley, John
Baddeley, Thomas
Blakeway, Charles
Hawkes, John Forrester
Morris, Thomas
Norris, T.
Taylor, Edward
Taylor, Joseph
Underhill, Thomas

Aston (Newport)
Norris, William

Berrington
Norton, Edward
Spenser, John

Bishopscastle
Bond, Henry Charles
Bullock, Gilbert
Edwards, Edward
Edwards, William
Griffiths, William Henry
Hay, Thomas
Jepson, Forrester
Marston, William
Matthews, John
Matthews, Richard
Matthews, Thomas
McNiece, John

Bridgnorth
Addison, John
Bickerton, Thomas Osten
Dyxson, Roger
Edwards, Samuel (1)

Edwards, Samuel (2)
Felton, George
Felton, Richard
Fisher, Stephen E.
Furber, Thomas
Gittos, William
Glase, Edward
Glase, Thomas
Green, George
Green, George & Son
Gregory, Richard
Harris, William
Hayward, Thomas
Higgons, Joseph
Hinksman, James
Hopwood, Robert
Lloyd, Richard
Newton, Isaac
Osborne, Joseph
Pearson, James Molesworth
Rogers, John
Stephens, Richard
Stokes, John
Street & Pyke
Street, Richard (1)
Street, Richard (2)
Taylor, William
Tailor, John
Wheeler, Cornelius
Woodward, Edmund John
Woodword, Maria

Broseley
Blakeway, Thomas
Hartshorne, William
Onions, W.
Reynolds, David

List of Clock and Watchmakers

Caynham
Hammonds, John

Cheswardine
Woolrich, Thomas

Church Stretton
Bond, Henry Charles
Moore, John
Stockwell, Thomas
Thomas, Richard

Cleobury Mortimer
Ames, Thomas
Newall, Henry
Newall, William
Palmer, William
Sheppard, George
Stockwell, Thomas
Young, –

Clun
Price, Thomas
Walters, William

Coalbrookdale
Fletcher, George

Cockshutt
Joyce, William

Condover
Darken, Mr.

Cotton (Wem)
Calcott, John (1)
Calcott, John (2)
Calcott, Richard
Grosvenor, John

Dawley
Banks, John
Burroughs, James
Burroughs, Rowland Kitson
Deakin, William
Edwards, Samuel
Seaman, Edward

Sterry, Robert
Wilson, William

Ellerdine
Wellings, William

Ellesmere
Armstrong, William
Banister, James
Bickerton, George
Bradshaw, George
Bullock, Edmund
Bullock, Jeremiah
Bullock, Richard
Bullock & Davies
Cartwright, William
Cross, William
Davies, Edward (1)
Davies, Edward (2)
Grosvenor, Robert, Edwin
Hunt, Robert
Joyce, John
Makin, –
Marsh, G. T.
Studley, Thomas
Taylor, Thomas

Hinstock
Wynn, William

Hodnet
Long, John

Hope
Sheen, William

Ironbridge
Burroughs, James
Burroughs, Rowland Kitson
Green, Walter
Greenfield, Joseph
Hinckley, William
Liseter (Lisellen), William
Peplow, Francis Young
Stafford, –
Webster, John

Ludlow
Ashby, John
Aston, Samuel
Brodhurst, Walter
Bryan, William
Burnett, Charles
Camell, William
Charlton, Edmund Lechmere
Clench, Richard
Crumpe, Richard
Daniell, Hugh
Edwards, Edward
Edwards, George
Edwards, Robert (1)
Edwards, Robert (2)
Edwards, Rowland
Edwards, William
Eston, Samuel
Farmer, Joseph
Felton, Richard
Griffiths, William Henry
Herbert, Thomas
Herbert & Son
Herbert, William
Higgs, Thomas
Jones, William
Jookes, Philip
Knight, Stephen
Langford, William
Leach, Edwin
Massey, Charles
Morris, John
Noke, James
Palmer, Joseph
Palmer, Thomas
Palmer, William
Payne, George (1)
Payne, George (2)
Payne, William
Percival, William
Phillips, Samuel
Phillips, Thomas
Phillips, William (1)
Phillips, William (2)
Powell, William
Roberts, Edward
Season, Thomas

Silvester, John
Stead, John
Thompson & Williams
Tipton, Benjamin
Turford, James
Turford, James Benjamin
Vernon, Thomas
Walker, Charles
Wood, Benjamin
Wood, Samuel

Madeley
Palmer, Elias Bailey
Peplow, Samuel Kirk
Pritchard, George

Madeley Wood
Russell, James

Maesbury (Oswestry)
Francis, Edward

Market Drayton
Arkinstall, Francis
Bowker, George
Bowker, John
Grosvenor, John
Grosvenor, Robert
Hulse, Ralph
James, David
Kaye, Joseph B.
McQuinn, John
Rodgers, George
Weatherby, Thomas

Minsterley
Everall, John
Lasseter, Henry

Much Wenlock
Blakeway, Thomas
Byn, *alias* Byrd or Burd, Richard
Corvehill, William
Haslip, William
Jeffries, Edward
Massey, John
Osborne, Joseph

List of Clock and Watchmakers

Thornton, Thomas H.
Waler, William
Wheeler, George
Wheeler, John

Newport
Baddeley, George
Barrett, John (1)
Barrett, John (2)
Basford, Daniel
Benbow, Thomas
Bowyer, William
Collier, -
Cooper, Thomas
Eccleshall, Charles
Jervis, Henry
Miller, Isaac
Morris, William
Norrison, William
Northwood, James
Orme, Michael
Simpson, Charles
Underhill, William
Whiston, Joseph
Whiston, Thomas

Northwood (Prees)
Benbow, John
Benbow, Thomas

Oakengates
Brown, Joseph
Edwards, -
Ellis, George
Morgan, Thomas
Vickers, George Henry

Oldbury
Fisher, John
Smith, John
Woller, Charles

Oswestry
Barnett, Jonathon
Birchall, Samuel
Bullock, Edward
Campbell, Francis

Campbell, Henry
Clarke, Elijah
Corken, Archibald
Evans, George Edward
Evans, Richard
Gardener, John
Giles, Henry
Hall, Robert
Higgs, William
Highfield, William
Hugh ap William
Jones, Joseph
Jones, Humphrey
Jones, Thomas
Lashmore, Edward
Lyons, Aaron
Matthews, James Howell
Matthews, Richard
Matthews, Thomas & Richard
Moody, John
Owen, William
Owen, William & Thomas
Payne, William
Phillips, Samuel
Roberto, William A.
Rogers, Joseph
Salter, Joseph
Salter, Robert
Smith, William ap Robert
Smythe, Robert
Stanton, Thomas
Thomas, Richard
Tomley, John
Turner, Henry C.
Wicksteed, Charles
Willoughby, Richard

Prees
Callcott, John
Edge, Griffith
Furber, Thomas
Houghton, Richard
Loton, John
Powell, John

Roddington
Brisbourn, Peter

Ellis, Samuel
Keeling, John

Rowton (High Ercall)
Hazeldine, William
Hazeldine, –

Rushbury
Blakeway, John
Blakeway, Thomas

Ruyton XI Towns
Brown, Thomas

St. George's (Oakengates)
Freeman, Thomas

Selattyn
Jared, Ann
Jared, William

Shifnal
Baddeley, Thomas
Blakeway, Charles
Chune, Thomas
Davis, John
Davis, Thomas
Davis, William
Davis, William, Henry
Loseley, Edward
Nicholds, Thomas
Peplow, William
Woolley, George

Shrewsbury
Anderton, William
Andrewes, Edmund
Aris, Philip
Aspinall, William
Baker, William
Barratt, Charles
Baxter, John
Baxter, Mary
Birchal, George
Bradshaw, Ellis
Brodshawe, Adam (1)
Brodshawe, Adam (2)

Brown, Henry
Campbell, Robert
Capper, John
Careswell, Francis
Carswell & Harper
Clevely, Thomas
Clifton, Cuthbert
Cooper, Joseph
Cotterill, Henry
Cross, William
Davies, Daniel
Davies, David
Davies, Henry
Davies, John
Davis, William (1)
Davis, William (2)
Doncaster, Edwin
Edmunds, John
Edwards, William
Evans, John (1)
Evans, John (2)
Evans, Mary
Evans, Philip Henry
Evans, Price James
Evans, Richard (1)
Evans, Richard (2)
Evans, William (1)
Evans, William (2)
Evans & Barnett
Evans & Brown
Fesser, Andrew
Fletcher, Charles
Fletcher, George
Gentry, John
Giles, Richard (1)
Giles, Richard (2)
Glover, George
Glover, John
Gorsuch, *alias* Gossage, Thomas
Gottlieb, Andrew William
Grant, J.
Green, John
Hanny, James
Hanny, William Stourton
Harley, Samuel
Harley & Son
Harley, William (1)

List of Clock and Watchmakers

Harley, William (2)
Harper, Richard
Harper, Thomas
Harper & Son
Harris, Richard
Harris, William
Hasty & Son
Hay, Cecilia
Hay, Thomas
Hay, Thomas William (1)
Hay, Thomas William (2)
Hays, Thomas
Hilton, Evan
Huber & Co.
Huber, Lawrence
Hutton, –
Ireland, John
Jackson, William
Jepson, Forrester
Jones, Henry Charles
Jones, James
Kelvey, Rebecca
Kelvey, James
Kent, John
King, James
King, Joseph
King, R.
King, Thomas
King, William
Last, William Bradbury
Launton, Thomas
Leigh, Richard
Macklin, Peter
Marston, William
Martin, William
Massey, John
Mew, Samuel
Middleton, John
Milligan, Thomas
Millington, Thomas & Co.
Morecock, Daniel
Morris, Richard & John
Morris, Robert
Nash, Richard
Nash, Thomas
Newnham, Samuel
Newton, Joseph

Nightingale, J. T.
Oswell, –
Pedley, Samuel
Pritchard, Thomas
Pugh, Benjamin
Rathbone, John
Robinson, Edward & Co.
Robinson, Henry
Robinson & Wells
Rossi, Joseph
Rowley, Henry (1)
Rowley, Henry (2)
Rowley, William
Savage, John
Savage, Richard
Savage, Thomas
Savage, William
Simpson, Christopher
Smith, John
Smith, William
Stone, Samuel
Stone & Allen
Super, William
Thomas, Richard
Thompson, Thomas
Traunter, Thomas
Trentham, Thomas
Walker, William (1)
Walker, William (2)
Warner, John
Warner, Rebecca
Webb, Joseph
Webster, James (1)
Webster, James (2)
Webster, John
Webster, John Baddeley
Webster, Robert
Welch, Henry
Wells, Henry
Wood, George
Wood, Isaac
Wood, James William
Wood, Mary
Wood, Richard
Woodruffe, James
Woodruffe, Thomas

Sutton
Wood, Samuel

Sutton Maddock
Hinksman, James

Tong
Baddeley, John
Ore, Thomas
Speight, James
Woolley, William

Wellington
Benbow, Thomas
Cetti, Paul & Co.
Clarke, Jane
Crispe, Edward M.
Davis, William
Dawson, William
Del Vecchio & Cetti
Del Vecchio & Dotti
Dotti, Dominic
Evans, William
Ford, Hannah
Giles, Richard
Gill, Caleb
Gutteridge, Job
Harris, Richard
Harvey, John
Holmes, Samuel
Huber & Fesson
Huber, Lawrence
Lawley, Joseph
Lawrence, Richard
Levi, Abraham
Peplow, William
Pitman, Arthur
Plimmer, Abraham
Plimmer, Nathaniel
Rivolta & Del Vecchio
Shaw, Joseph
Symon
Vickers, George Henry
Webb, William

Wilson, Henry
Winter, Samson
Wright, Thomas

Wem
Brown, John
Butler, Henry
Calcott, John
Cartwright, Thomas
Freeman, Edwin
Hill, Thomas
Pritchard, John
Ray, Samuel

Westbury
Rider, Job

Whitchurch
Bradshaw, George
Bradshaw, Joseph
Calcott, John
Churton, Joseph (1)
Churton, Joseph (2)
Cooper, Benjamin
Cooper, Charles
Cooper, Joseph
Cooper, Mr.
Deaves, Richard
Finn, Thomas
Henshall, Henry
Jarvis, John
Joyce, James (1)
Joyce, James (2)
Joyce, John Barnett
Joyce, Thomas
Joyce, Thomas & Son
Lamb, John
Murray, Joseph
Newnes, Samuel
Ridgway, Josiah
Torkington, Jeffrey

Wootton (Oswestry)
Bullock, William

List of Clock and Watchmakers

BAROMETER MAKERS AND RETAILERS

Bishopscastle
Fosanelli, Peter

Coalbrookdale
Sankey, J.

Cotton (Wem)
Callcott, John

Ludlow
Laffrancho, J.

Market Drayton
McQuinn, John

Oswestry
Owen, William
Pozzi, Peter

Shifnal
Davis, William

Shrewsbury
Baker, William
Bowley, William
Davis, E.
Gianna, Lewis
Gittins, W.
Jacopi, C.
Lombardini & Casteletti
Pedroni, J. B.
Rossi, Joseph

Wellington
Del Vecchio & Dotti

ADDENDA

Del Vecchio Wellington, 1857
Del Vecchio.—21st May, after a brief illness, aged 62, Mr. Gaetano Del Vecchio, jeweller and furniture dealer, of New Street, Wellington; a native of Laglio, Lake of Como, Lombardy. (*E.J.* 3 June 1857.)

Griffiths, George Middlewood, 1861
20th June, much respected. Mr. George Griffiths, watchmaker, of MiddleWood in this county. (*E.J.*, 26 June 1861.)

Mason, Joseph Wellington, 1875
Joseph Mason, watchmaker, of Wellington, charged Jabez Grosvenor, ironmonger, of New Church Road, with an assault. (*E.J.*, 23 June 1875.)

Powell, William Ludlow, 1859
Powell-Crundall-15th May, at St. Lawrence's Church, Mr. William Powell, jeweller and watchmaker, to Miss Emily Crundall, both of Ludlow. (*E.J.*, 18 May 1859.)

Savage, John Shrewsbury, 1848
'On the 15th inst. after a severe illness, Mr. John Savage, of the Peacock Inn, Abbey Foregate, formerly clock and watchmaker, of this town.' (*E.J.*, 16 August 1848.)